The Authors

John Rose is manager of Broadstone Stud, newly set up to breed competition horses. He was stud manager at Catherston Stud and the Warwickshire College of Agriculture where he successfully bred and showed in-hand hunters and lectured in Stud Management. This wealth of practical experience, including seven years running an international dressage yard in France, has been combined with knowledge of equine physiology of *Sarah Pilliner*, formerly Lecturer in Equine Science at the Warwickshire College of Agriculture. She is the author of *Getting Horses Fit* and both authors are regular contributors to the popular equine journals.

Practical
Stud Management

John Rose
and
Sarah Pilliner

BSP PROFESSIONAL BOOKS

OXFORD LONDON EDINBURGH

BOSTON MELBOURNE

First published 1989

British Library
Cataloguing in Publication Data
Rose, John
    Practical stud management.
    1. Livestock: Horses. Breeding.
    Stud farms
    I. Title   II. Pilliner, Sarah
    636.1'082

ISBN 0−632−02031−8

BSP Professional Books
A division of Blackwell Scientific
    Publications Ltd
Editorial Offices:
Osney Mead, Oxford OX2 0EL
    (Orders: Tel. 0865 240201)
8 John Street, London WC1N 2ES
23 Ainslie Place, Edinburgh EH3 6AJ
3 Cambridge Center, Suite 208,
    Cambridge MA 02142, USA
107 Barry Street, Carlton, Victoria 3053,
    Australia

Set by Setrite Typesetters Ltd
Printed and bound in Great Britain

Dedication

This book is dedicated to Cedola, the foundation brood mare which was a gift from our Dutch friend Mrs Benedictus, with thanks for the many champions she produced for so many people.

Contents

Foreword

I was delighted to be asked to write this foreword as this is a book that is much needed. It is an extremely practical and straightforward book answering very clearly all the questions one is so often asked by the mare owner. It gives the stallion owner help on all aspects of stud management, sets a good standard for handling stallions, covering, foaling and explains simply any problems which one is likely to face when running a stud.

The section on Artificial Insemination and Embryo Transfer is a very necessary subject and one which is causing tremendous interest in this country at the moment. It is so important that Britain moves with the times and becomes really proficient in equine breeding.

It would be difficult to find a book with so much clear concise detail on this subject and anyone considering running a stud or pursuing a career with horses will find it an invaluable guide. Practical Stud Management will be a useful source of reference for any stallion or mare owner wishing to breed a fit and healthy foal.

Jennie Loriston-Clarke MBE FBHS

Acknowledgements

Our thanks go to Mrs. Joan Pendlebery for her hours spent deciphering our illegible writing and coping with the 'naughty bits'; Clive Milkins and Kevin Randall for the photographs; Dawn Rose for moral support (and coffee) and Jeremy Houghton Brown for making us do it in the first place!

Introduction

Horses and ponies breed successfully left to their own devices in a natural environment, so why should a book such as this be necessary?

First, twentieth century horses do not live in a 'natural environment'. We take mares and stallions, separate them and then re-introduce them very briefly during a breeding season decided by us. Consequently the mare has a lower conception rate than ewes, cows and sows.

Second, horses have been selected for performance, not reproductive efficiency; winning races or jumping fences is more important than breeding foals, and faults such as a tendency to breed twins have been allowed to continue.

Finally, successful competition horses are very valuable and stud fees are high. Thus the horse breeding industry needs to be efficient and give value for money, and owners need to appreciate the problems involved.

The aim of this book is to examine the breeding of competition horses from two points of view; that of the mare owner and that of the stud manager. There is no reason why sending a mare to stud should be shrouded in mystery; the mare owner has an important role to play in preparing the mare before she goes to stud so that she is in optimum condition for breeding and subsequently rearing the foal. It is increasingly important that stud managers maintain high standards of stud and stable management and also keep abreast of the modern veterinary and feeding techniques outlined in this book.

The first consideration of a mare owner must be what type of foal to breed. In the UK we are privileged to have a wealth of different blood lines to choose from; not only our racing thoroughbreds but also outstanding native breeds – the Irish Draught and the now well-established British Warmblood as well as the foreign horses standing in this country.

The mare owner must also consider just what is involved after sending a mare to stud. The stud manager takes over; it is his or her job to try to get the mare in foal. A sound understanding of the physiology behind the process is needed as well as practical stud and stable management. Understanding becomes even more important as artificial insemination, frozen semen and embryo transfer become more commonplace.

Both parties need to be familiar with the process of foaling and how to deal with any problems that arise. The book also deals with nutrition, infertility, parasite control and stud design.

Although this book could never replace the years of valuable experience achieved by top breeders, the authors hope it will be of assistance to many people wishing to further their knowledge of horse breeding.

1 What type of foal to breed

It is important when breeding a foal to decide first the type of animal you wish to breed. For example, you will not produce a point-to-point horse by crossing a Welsh pony with a Shire!

The dressage horse

Let us take the dressage horse into consideration: strength, movement and temperament are the first criteria. Strength is especially important in the back and hocks. The horse should be an excellent mover with good, supple paces and plenty of 'swing' to the trot and should show some signs of natural extension at this pace. Nevertheless, it is not always the horse with the 'flashy' trot that will produce the best lengthened strides and extensions which are so important these days in any level of competition, be it novice or Grand Prix. The horse should also have some bend to the knee, so that advanced movements such as Passage and Piaffe will be easier to achieve.

Although desirable, many top dressage horses of world class fame are not always perfectly straight in movement, but the ability to remain balanced at all times at trot is vital, even when ridden around corners. The walk must not be neglected as many valuable points can be lost from a bad walk. The points to look for are a good four-time movement which is not hurried. The canter should be equally brilliant and all paces should be naturally balanced and light.

Temperament should depend upon the level required as it is useless to have a horse which lacks courage in the hope that it will conquer the difficult movements. A 'gassy fizzy' temperament, however, is not desirable either.

As already stated, strength is of great importance, as many hours of work are required to school a horse to the highest level in

Fig. 1.1 What type of foal to breed is the first consideration.

dressage; if a horse has the suitable movement and temperament, these hours of work will doubtless be shortened, thus prolonging the horse's competitive life.

Good looks are not always necessary when wishing to impress the judge, although bad conformation may alter the brilliance of paces. There is a growing trend to compete stallions for several reasons; one being that they are intelligent and can cope with the work load; another reason is the financial rewards of standing a successful stallion at stud. These animals should have good looks and good conformation if a wider market is aimed for. A dressage stallion is capable of siring hunters, event horses and show jumpers as well as dressage horses. A good-looking horse which can perform is a bonus to any owner.

Any combination of breeds can produce the dressage horse but generally a smaller percentage of thoroughbred blood is desirable.

The Warmblood is very fashionable at present, as their temperament, strength and movement are generally good. Thoroughbreds, pure Arabs and cross-breds have, however, reached international level.

The show jumper

The show jumper's requirements are not unlike those of the dressage horse. In fact, many knowledgeable people say that a good dressage horse should jump well and vice versa. A much 'rounder' trot is required but the canter should be equally as brilliant as that of a dressage horse.

When choosing a jumping sire it is desirable to find one that has either proved himself or produced a number of good jumpers from different types of mare. By comparison, a show jumping horse must have natural talent to reach top-class competition level, but a dressage rider can improve his horse far more with training. No trainer can force a horse to jump if the horse does not possess that certain natural jumping ability. Above all, the horse must be so careful as not to want to touch the poles with his feet or limbs.

Most combinations of breed can produce a show jumper including, of course, the Thoroughbred, Warmblood, Irish Draught, Cleveland Bay, Shire and native breeds. Certain countries appear to favour certain breeds; for example, in the 1970s, the Americans used thoroughbreds, and more recently the German breeds have become more popular.

The French have their own breeds, the Selle Francais and Anglo-Arab 'Francais'. The Germans have their own breeds of Warmbloods — Hanoverian, Holsteiner, Oldenburg and Trakener to name a few. The Dutch, Swedes and Danes also have their own Warmbloods and the Italians in the past have favoured the Irish Draught crosses. Great Britain does not appear to favour a certain breed, but at present the fashion remains to 'go abroad'.

The three day event horse

The event horse requires great stamina and speed, therefore the thoroughbred plays an important role. Nevertheless, many famous event horses have a small percentage of other breeds in their pedigree; some say a touch of pony blood will improve the animals' agility when jumping. These days, a tough sound horse that moves well is required, particularly as the dressage phase plays an increasingly important part. Many events have been lost in the final day's showjumping because the previous day the horse has been asked to cover several miles of roads and tracks, a fast steeple-chase phase and then the cross-country course, which in itself is a very demanding section. Therefore, to jump a clear round on the third day requires a courageous horse that is sound and nimble. Most horses have the ability to complete a novice course, but it takes a very experienced horseman to select a potentially top-class three day event horse.

Pleasure or profit?

Whichever type of animal one decides to breed, be it a show pony or race horse, finance is a key consideration. Are you breeding for pleasure or to make a profit? Generally, there is always money to be made in thoroughbred breeding, although proven broodmares and stud fees are extremely expensive, and in most cases a large financial outlay is required initially. When a market for a certain type of horse is recognised, this market will soon become flooded. It has happened in the past in the pony and thoroughbred breeding industries where inferior types of horses and ponies have been bred, creating a large number of animals which have not 'made the grade'. However, there is always a good market for any animal of the highest quality, whatever the breed and, to be realistic, an animal of poor quality is destined for the meat trade.

There is a worldwide market for top class competition horses either to produce and sell foals or to run on the youngstock to break and sell properly produced at three, four and five years old. This appears to be where the profit lies, if the owner/breeder has adequate facilities and grazing.

Whichever type of horse you decide to breed, the basic principles apply to each one, except that the thoroughbred appears to be more difficult to breed from than, for example, the native pony. Breeding equines is not as easy as it looks and many unprofessional people expect instant results when sending their mare to stud. It is extremely satisfying for a stud or stallion owner if good results appear quickly, thus enhancing his or her profit and reputation.

Although you may not be aiming to make a profit, be absolutely sure that your mare is correct and sound − it is no pleasure to breed an incorrect foal.

Choosing the mare

When choosing a good broodmare, certain factors should be taken into consideration. Although a good stallion can improve a poor mare, she will contribute at least fifty per cent towards the eventual foal's qualities and some mares consistently produce top-class foals. Therefore, only mares of proven breeding ability or mares which have proved themselves either in the show ring, competition world or racing, should be bred from. It must also be remembered that it is not only the top class mare which will breed winners; many top racing fillies, for example, which eventually retire to stud never breed winners themselves, although this may not be the case in subsequent generations. In other words, a good pedigree is also important. Similarly, in the showing world, it is not always the champion mare who produces the champion foal. Mares which have 'broken down' or have a family history of unsoundness should not be bred from. Remember, there is always an element of luck in horse breeding and some excellent horses have been bred from stock possessing unlikely parentage.

Age

Although some of the 'heavier' types of mare have successfully been bred from at the age of two years it is unwise to breed from a mare under the age of three years. It must be remembered that young horses continue to grow until they are at least five years old and sometimes older; to breed from a young filly could affect her own maturity. If one does consider breeding from a three year old filly, then she must of course be fed accordingly and must only be bred from if she is sufficiently mature and advanced for her age. If bred from while she is too young, the filly may not conceive as the ovaries may not be sufficiently active. The safe maximum age to breed from a mare is difficult to advise; mares have successfully bred incredibly late in life, some until 35 years old, while others may 'give up' from 15 years onwards or before. As with a young mare, it really depends upon her health and physical condition. It can be an advantage if the mare has had a foal earlier on in life as this sometimes helps fertility later, that is to say when her competitive or racing career has ended and she has been retired.

It is worth taking note that a mare's temperament may be affected after having had a foal, especially in the case of a young mare. Should she be required to compete later in life she may become more independent and more reluctant to accept discipline, and difficult to train.

Conformation

It is generally known and accepted that it is not always mares with the best or most correct conformation that breed the best foals. Nevertheless, when breeding from a mare for the first time, her conformation is a very important factor and must be taken into consideration when it comes to choosing a suitable stallion.

A 'maiden' mare tends to breed a small foal. It must be remembered that the mare with the most sloping shoulder does not always move better than one with a straighter shoulder. If, for example, your intention is to breed a good hunter or eventer, the forehand has a great bearing on the balance of the rider when jumping drop fences. Where a mare is being used for show purposes she must, of course, have excellent conformation, whatever the breed, and she must be true to type. Although, technically, the head is of least importance, an oversized heavy plain head may affect a horse when being schooled for dressage. The head should be well 'set on' to the neck, supported by a good sloping shoulder. The mare should have a good depth through the girth, a correct foreleg with plenty of good bone, short cannon bones and strong round feet (contracted heels are much frowned upon). Bone is, of course, the all important factor when choosing a mare or stallion.

The mare should not be too long in the back, though generally broodmares should be 'roomy'. She should have a good strong loin and quarters with a well 'set on' tail. Equally, it is important that her hind leg is not too straight or too bent, as this will affect the youngster's training should it inherit these faults.

When describing the mare's conformation, one must understand the importance of the shape of her vulva. It should not be too sloping or long, which is common in older mares as shown in Fig. 1.2. This can only affect her subsequent fertility, due to the

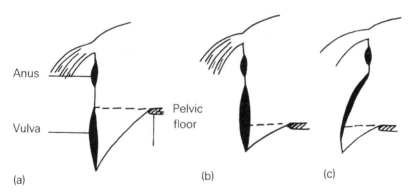

Anus

Vulva

Pelvic floor

(a)   (b)   (c)

Fig. 1.2 Vulval conformation. (a) Ideal: most of vulva below pelvic floor. (b) Poor: most of vulva above pelvic floor; area above dotted line to be stitched. (c) Poor: tilted vulva allowing air into tract; needs stitching above dotted line.

possibility of dung passing into the vulva, thus creating infection. It may be possible that air will be taken in during service; this, too, will create infection and infertility and is known as pneumovaginitis. This defect can be rectified by a simple 'Caslick's' or stitching operation carried out by a veterinary surgeon after service (see Fig. 1.3). Grey mares may suffer from melanomas (black cancer). These are lumps often found under the skin around the head and the dock which could possibly affect her breeding potential. An internal examination by the veterinary surgeon is advisable, especially if any such problems are suspected, before she goes to stud.

Hereditary defects

A number of hereditary defects should be avoided, including parrot mouth or overshot jaw (Fig. 1.4), where the top set of teeth protrudes beyond the bottom set, thus making it difficult for the animal to graze successfully. This should be noticed immediately at birth and in severe cases, where the teeth pass by several centimetres, the foal should be destroyed immediately as it will not be able to thrive successfully. Should the top teeth only touch the bottom set or not pass over, this is not generally accepted as being parrot mouthed.

A mare suffering from navicular disease, if hereditary, should not be bred from. However, if she contracted this disease of the

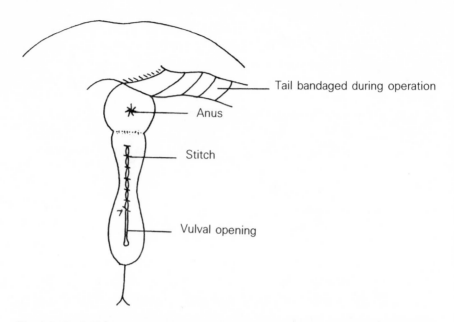

Fig. 1.3 Caslick's operation. This is a relatively simple procedure which prevents infection entering the vagina and subsequently the uterus. The vulval lips are stitched together to a point below the pelvic brim. The veterinary surgeon injects local anaesthetic into the vulval lips, trims them to form raw edges and then stitches these edges together. After the area heals the stitches may be removed and a seal is formed which prevents air and dung entering the vagina. A small opening at the lower V-shaped portion of the vulva allows urine to escape.

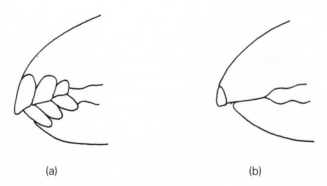

(a)   (b)

Fig. 1.4 (a) Parrot mouth: adult horse. (b) Severe parrot mouth: foal.

foot after many years of hard work on hard ground then she must be considered as a 'sound' mare suitable for breeding. Other defects are 'string halt' (less common these days), bone spavin, ringbone, sidebone, roaring, whistling, cataracts and shivering. Cribbing is not strictly an hereditary defect but should a mare that crib-bites be bred from, this vice will easily be copied by her foal and any other foals in her company.

Temperament

Temperament is sometimes difficult to assess; a broodmare, as in any other successful horse, must show a certain amount of tough character and it will be an advantage if she possesses this attribute. This will undoubtedly be passed on to her offspring and help in the production of suitable and successful competition horses. Some of the most outstanding horses in all spheres of competition have had, to say the least, a strong character. Mares with a difficult temperament should not be bred from as they may pass this on to their progeny. Equally, the mare should not be too placid; this temperament may suit some top horses and the majority of this type of animal may be suitable for the novice rider, but when asked to jump more demanding fences and travel at greater speed you could possibly find the horse just as unsuitable as a horse with a 'difficult' temperament.

Pedigree and performance

It is advisable to know the mare's breeding or pedigree so that the important bloodlines can be traced and this will also affect the ultimate value of the offspring produced. There are some families of horses which continually breed success in many areas of the horse world and these should be carefully noted when choosing a mare. The performance record of the mare is important and it is an advantage if she is proven competitively. An alternative would be to breed from the full-sister or close relative of a gelding or stallion competing or racing successfully.

The mare should ultimately be a good mother; there is no point in breeding from a mare that will not accept her young or not produce enough milk to enable the foal to develop to its best advantage.

Choosing the stallion

It has often been said that 'choosing the stud' where the stallion stands is of prior importance as owners with top-class mares are naturally anxious and particular about the way in which their mares will be cared for. Mares may be injured due to inferior stud management, although it must be accepted that an accident may occur at home as easily as when the mare is away at stud. It is all too easy these days to seek legal advice for something that may not be the fault of the stud owner or his staff. There are many ways of finding a suitable stallion; for example, there are equestrian publications which advertise stallions at stud under their various breed sections. There are stallion directories for thoroughbred and non-thoroughbred horse which are published nationally. If necessary, write to the various breed societies for the relevant information on their approved stallions.

Since National Stallion licensing was discontinued most breeds are now approved by their various societies. Word of mouth still remains one of the most reassuring ways of finding a good stallion. The mare owner must be prepared to travel if necessary, although this adds quite considerably to the expense and the time the mare will stay at stud. Like the mare, it is an advantage if the stallion has a proven racing or performance record and has retired sound. In an older stallion it is important that his progeny are also doing well. When breeding any type of horse, quality and quantity of bone is essential as it is so easy to breed-out bone in as little as one generation if the wrong stallion is chosen for a certain mare. A competition stallion should have at least nine inches of bone, even if the mare has adequate bone herself. A stallion with less bone is acceptable only when breeding from a very heavy mare. The stallion's conformation must be considered and should certainly complement that of the mare. Ideally, a mare with a 'long back' should be put to a stallion with a short one. A plain head can be improved by using a stallion with an attractive head but cross-breeding can be unpredictable. The movement should be outstanding for its type or breed and presence should be in abundance, although one must take into consideration the stallion's age; it could be that an older stallion will not move as fluently as he did when he was younger.

A stallion should look like a stallion and not like a mare or gelding; he should have a suitable temperament but it must be understood that an entire during the covering season will generally be considerably more difficult to handle, in comparison with the autumn and winter months.

A stallion with 100 per cent proven fertility is very rare; however, one should use a horse with as high a fertility percentage as possible. Remember that the percentage is concluded from the amount of mares covered and which are subsequently in foal; most studs have some non-breeding mares which may unfairly affect the stallion's percentage rate. The stallion should also be free from the same hereditary defects as the broodmare.

Viewing the stallion

Having selected several suitable horses to look at the mare owner should send for stud cards. A stud card gives a detailed description of the stallion and his performance record, stud fee and keep charges, and details of his winning progeny, etc. A well printed stud card with a good photograph reflects the quality of the stallion. The next step is to make an appointment to see the chosen stallions, and if a route is mapped out several stallions can be seen in one day.

First impressions are very important and on arrival at the stud the general layout, fencing and grazing should all be carefully observed. The stallion should first of all be viewed in his box where it is easier to assess his temperament. Here again, your first impression is important and will affect your final decision; he should 'catch your eye' and your impression of him must improve the more you look at him rather than the reverse effect!

The level of stable management should be assessed and the stallion must be in prime condition: not too fat or too thin, especially at the beginning of the covering season. He should then be viewed outside and seen trotted up in-hand by which time he should have shown you several signs of brilliance, i.e. conformation, movement and presence. In the case of the competition stallion which competes on a regular basis, it may be possible to see him at work or even jumped. Some studs have their own Viewing and Open Days where the stallions can be seen working and their young stock are on show.

The stock should be viewed whenever possible; ask to see the mothers of the youngstock to compare their physical condition or faults in conformation. A good stallion should consistently produce a certain type of stock from all types of mare. However, if you discover some common faults in his offspring such as straight hocks, curbs or common heads, then he should be avoided. Colour, too, is a factor to be considered, as certain colours are fashionable when it comes to selling. Some stallions do not always throw a consistent colour but some do, giving prospective breeders an idea of what may be expected from their mare. The physical condition of mares and stock must be observed though a mare in a thin or poor condition may just have arrived at the stud. Check the condition of the mare's feet and ask if visiting mares are wormed regularly.

If you would like your mare to foal at the stud, ask to see the stabling and foaling facilities and chat to the staff where possible. If they find the time to speak to you and answer your questions, then usually you will find they will have the time to look after your mare correctly.

Stud fees and terms

Ask for stud fees and keep charges.

The example invoice shows an account likely to be incurred for one mare while at grass only. The amount shown would be considerably more should you wish the mare to be stabled.

Example invoice

Stud fee	£500
Groom's fee	£5
74 days keep at grass @ £2 per day from 1 May until 14 July	£269
Worming	
18 May	£6
18 June	£6
Foaling fee	£50
Blacksmith trim 29 June	£6
	£842
+ VAT @ 15%	£126.30
TOTAL	£968.30

Other expenses likely to occur are: transport to and from the stud; insurance; veterinary charges; keep of mare while in foal and up until foaling.

Methods of payment

There are various conditions of payment of the stud fee. For example:

Straight fee

This is where the fee is payable as soon as the mare has been covered and is paid irrespective of whether the mare is tested in foal or not. This usually applies to the lower stud fee bracket.

No foal – no fee: 1 October terms

The stud fee is payable once the mare has been covered and if the mare is subsequently certified barren on 1 October, then the stud fee only is returned. This is generally favourably accepted by mare owners.

No foal – free return: 1 October terms

Here again the stud fee is payable once the mare has been covered but, should the mare be certified barren on 1 October, a free service is given the following season only. These terms are generally accepted by most mare owners, especially if the stallion is young and his stud fee will be increased the following year. Using this method the stud owner will find it a little easier to budget, but naturally must ensure that the stallion will be standing at stud the following season. Should the stallion meet with an accident, die or be sold, then the stud owner must be helpful in finding a substitute for any owners with barren mares.

Part payment

Part payment usually applies to the higher stud fee bracket. Half or part payment is paid after the mare has been covered and the rest on 1 October, if the mare has been certified in foal.

Live foal

The stud fee or part stud fee is returned should the mare not have a live foal the following year. This normally applies should the foal not live for more than 48 hours and again only applies where stud fees are extremely high.

Concessions

One may often see concessions to winners and dams of winners. This does not only apply in thoroughbred breeding: mares that have been successfully competing in showing, eventing, dressage and showjumping are often granted a concession.

The stallion owner will encourage good mares to his young stallion in his first season standing at stud; a reduction in stud fee is occasionally made and exceptional mares are sometimes given a free service.

2 Sending a mare to stud

It is important for the mare owner to be thoroughly organised before sending the mare away to stud.

Nomination forms

Once the mare owner has decided to put the mare to a chosen stallion, a nomination form must be completed. This, as the name suggests, is an agreement to reserve a 'nomination' between the mare owner and stud or stallion owner. This may vary from stud to stud but the basic requirements are the same. Figure 2.1 shows an example of the general conditions of the agreement. Figure 12.6 in Chapter 12 sets out the other details required in a nomination form.

Some studs require a booking fee or deposit upon the completion of this agreement. The stallion owner may be expecting a certain number of nominations and can be easily let down for various reasons. Most studs require mares to be up-to-date with influenza and tetanus vaccinations and in most cases hind shoes are removed.

The completed nomination form should be returned as soon as possible to the stud, and notification of acceptance returned to the mare owner, especially in the case of 'approved mares only'.

Swabbing

It is important to 'swab' a mare before sending her to stud. With a maiden or barren mare the swab required is taken from the clitoral fossa for CEM (contagious equine metritis) and it takes approximately a week to obtain the result.

A second swab should be taken when the mare first comes into season. This is generally taken as a precaution to find out if the

General Conditions

1. Please make sure that mares arrive at stud without hind shoes.

2. *All* mares will be wormed at arrival and at regular intervals.

3. Veterinary assistance and advice will be obtained at the discretion of the staff on duty at the time. This will be chargeable to the owner.

4. The farrier will be called at regular intervals whilst the mare is at stud and will be chargeable to the owner.

5. Mares will only be covered on receipt of a certificate to confirm the negative results of a clitoral swab test for Contagious Equine Metritis (mare need not be in season). A further cervical swab will be taken on the mare's first day of season.

6. Mares will be accepted only with an up-to-date Flu/Tet & C.E.M. certificate.

7. The account for stud fees, keep fees and any other will be presented when the mare is collected and must be paid for in full at that time. V.A.T. at 15%/the appropriate rate is charged on all accounts.

8. No foal, no fee 1st Oct concessions are offered.
    a) The stud fee must be paid at time of collection of mare.
    b) Should the mare be barren on the 1st October, a veterinary certificate must be produced to prove this.
    c) On production of this certificate, the stud fee + V.A.T. only will be refunded.

9. Cancellation Fees. 25% Cancellation fee will be charged for any Nomination not taken up unless a Veterinary Certificate is produced stating that there is a valid reason for the mare not to be served.

10. Every care will be taken whilst mares are visiting, but no responsibility will be accepted for accident, theft or disease.

Fig. 2.1 General conditions set out in the first part of a Nomination Form. (Courtesy Broadstone Stud, Oxfordshire.) The details of the rest of the form are shown in Fig. 12.6.

mare has a uterine infection or in some cases when the mare looks 'dirty'. 'Dirty' mares are more obvious when in oestrus (in season) where a 'pussy' discharge will be apparent sticking to the tail and vulva.

These signs may indicate a vaginal or uterine infection and should covering take place, the infection will not only reduce the sperm life but the fertilised egg will not be able to survive in these conditions; thus the mare would become barren. The cervical swab is taken using a speculum in the mare's vagina so that the swab can be taken from the cervix (Fig. 2.2).

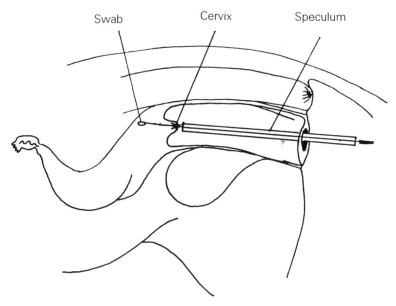

Swab   Cervix   Speculum

Fig. 2.2 Uterine cultures can be obtained by inserting a culture swab through the mare's cervix via a speculum.

Bacteria which can cause uterine infections are: *Klebsiella, Pseudomonas, coliforms, staphylococci, streptococci* and others which may cause abortion. These must be treated by means of irrigation or washing out the mare's uterus by the veterinary surgeon along with antibiotics. She will then be reswabbed the following heat before service may be allowed. Should a mare be covered by a stallion when she is dirty, not only will she not maintain her own pregnancy but the infection may be passed from mare to mare via the stallion. This could be disastrous, causing infertility of many mares and possible closure of the stud.

The foaling heat

Whether the mare foals at home or at stud, a decision must be made about having her covered on the 'foaling heat'. The rate of conception at this stage is approximately 50 per cent. Veterinary surgeons tend to disapprove of covering at this stage as, in some cases, the uterus has not had the time to retract to its normal size. Some mares take longer to 'clean up' after foaling and develop an infection; others may not be producing a 'ripe' follicle or egg ready to be fertilised.

After foaling, a mare generally comes into season one week later, although it is not uncommon for some mares to come into season as early as the fourth day after foaling or as late as two weeks after foaling. In some cases a mare may not appear to come into season at all!

Foals destined for a career on the race track or in the show ring need to be born as early as possible in the year, between February and April. If a mare foals rather late in June or July, it could be an advantage to use the foaling heat to bring forward the foaling date for the following year. Should the mare conceive and foal on time, several weeks would be gained. If the mare foals very late in July or August it may be better to leave her barren for that year and cover her early the following year. Having said the mare will come into season approximately one week after foaling, from then on she should 'cycle' normally and come into season every three weeks. If the mare foaled at stud and it has been decided not to cover on the foaling heat, a calculation for at least an extra three weeks' keep must be allowed for in addition to the normal charges.

It is possible to inject mares with prostaglandin approximately five days after she has 'gone off' or gone out of season to bring her into oestrus earlier and to help her regain some of the time which has been lost. The hormones involved in the oestrous cycle will be discussed in detail later.

Physical condition of mare

It is very important that mares should be sent to stud in a suitable physical condition for breeding. Mares in very poor condition are less likely to conceive. A higher level of nutrition is necessary to enable the mare to produce healthy follicles, but on the other hand, overweight mares may not conceive as too much body fat can cause cystic ovaries. A maiden mare is more likely to get in foal if she is on a rising or upward plane of nutrition and it has been proved on numerous occasions that a mare in 'light' condition brought into a stable and given good hay and concentrates will be more likely to conceive.

Taking a mare to stud

First of all, the owner must give the stud notice of his arrival. The stud manager has already had prior warning by means of the completed nomination form and most studs are 'geared up' to receiving mares daily. However, should the mare require to be stabled, an adequate warning is needed, especially if she is being sent to foal. If she is to be a grass livery, then shorter notice will be adequate.

It is also better to take the mare a few days before she is due to come into season, as she may need time to settle into her new surroundings. If the owner is not sure, or if the mare hasn't appeared to have 'shown', then she should be taken as soon as possible. It may be that she will only 'show' to a stallion or that she needs veterinary attention.

Worming

The mare should always be given a worming dose before sending her to stud, regardless of whether she will be wormed upon arrival. These days wormers are very easy and safe to administer and allow for overdosing. It is better to have her wormed twice running than not at all, especially when sharing pasture with many other strange horses.

Shoes

Generally, most studs insist that hind shoes must be removed. Ideally this should be done by the mare owner's farrier before arrival at stud as it is illegal for an unqualified farrier or stud assistant to interfere with the shoeing or trimming of the feet of any horse or pony. If the owner sends the mare with feet in good condition then she will most likely be returned in the same condition.

Tack

A stud will usually prefer owners to take home all tack, head collars, rugs, bandages and rollers. The less you leave with your

mare, the less likely it is to be mislaid or lost! Only in certain
instances where the mare is still in work or not properly 'roughed-
off' should rugs be left. These should be clearly marked and a
check-list made in duplicate.

Certificates

It is important to remember to send or take any swab certificates
and flu and tetanus identification cards to the stud with the mare.
    Finally, owners should try not to interfere or worry too much. If
the stud has been chosen correctly, the mare will be in good hands;
the risk of accident or disease will be no greater than if she were
still at home.

3 Getting a mare in foal

A 'normal' broodmare will come into season every three weeks.
She should arrive at the stud with negative CEM (contagious equine metritis) swab results as well as up-to-date influenza and tetanus inoculation certificates. If the weather is good and the grass keep of suitable quality, the mare is better off kept at grass; she will be more relaxed, come into season regularly and therefore be more fertile. However, with the current trend for early foals, be it for racing or showing, the breeding season commences much earlier in the year and it is necessary, especially in countries with an unpredictable climate, to keep the mares stabled and even rugged-up. It will also improve the chances of the mare coming into season early in the year by giving her seven to eight hours of 'daylight' from Christmas onwards. In other words, leave the lights on in the stable from 4 pm until about 11 pm; an easier and more practical method would be to install a time switch. The management of mares in this way helps to persuade the mare that it's already Spring and combined with a higher plane of nutrition this should encourage her to come into season earlier than normal, as the light helps to activate the hormone production in the pituitary gland. Whichever way it is decided that the mare will be kept, the teasing and trying routine remains the same.

Teasing or trying the broodmare

Under natural conditions where the stallion runs free with his mares, it is he that assesses when the mare is in oestrus and receptive to his attentions. In most studs, however, stallions and mares are kept separately, due largely to the high value of stock, even in the non-thoroughbred competition horse industry, and it is very important that the mare is completely receptive to the stallion before covering is attempted. This is one reason why we tease and try mares.

Normally, a mare needs to be 'tried' and covered every 48 hours because she ovulates towards the end of the time she is showing. Thus it will lead to over-use of the stallion if she is covered on the first day; she should be covered every second day until she is no longer showing.

It is accepted that sperm remain viable in the mare's reproductive tract for about 48 hours, and that providing there are no other problems the mare should conceive if she ovulates during this period. In some cases mares have conceived from a single covering even though they have subsequently remained in season for several days. This indicates that sperm life can be considerably longer that the accepted norm of two days.

Some unpredictable mares may only 'show' in season for a very short time; some only for as little as a few hours. If the mare is known to behave in this manner and still be fertile, it would be advisable to tease her every day. Mares may behave in this way at the 'foaling heat' which is when the mare first comes into season approximately one week after foaling and from then on every three weeks. Most veterinary surgeons do not recommend that mares are covered during this heat for two reasons: first the uterus may not have returned to its normal shape and second, she may not yet have 'cleaned up' after foaling and possibly be infected in some way. However, approximately 50 per cent of mares covered on the foaling heat will conceive. Other owners prefer to wait four or five days after the mare has 'gone off' and inject her with prostaglandin so that the mare comes back into season several days later and this way a high fertility rate can be obtained.

Rectal palpation

Nervous mares, if overteased, may become more nervous and, although producing healthy follicles, may not 'show' to the stallion at all. Some mares 'show' for several days yet may still not be producing follicles and if covered will not become pregnant. In these cases, regular rectal palpations by the vet will assist the stud groom to decide whether or not to cover the mare. A rectal palpation is an internal examination made by the vet inserting his hand into the rectum where he can feel through the wall of the rectum and can 'pick up' the ovaries to feel the size of the follicles they contain. A follicle is a fluid-filled sac containing an egg. A large follicle may be described as 'going soft' which means she is about to ovulate; this would therefore be a good time to 'cover' the mare. He may inspect her cervix by means of a speculum to see if it is open or closed; if it is open she is generally in season, the open cervix allowing the sperm to pass through into the uterus during covering. If the cervix is closed up and tight this is a sign that the mare is not in season.

In a busy stud covering only if the mare has a 'soft' follicle and is ready to ovulate will help to 'save' the stallion, thus keeping his fertility rate as high as possible throughout the stud season. In most thoroughbred studs or studs with busy stallions and high stud fees mares are regularly palpated and the mare is only covered when ovulation is imminent but an experienced stud groom will be able to reduce the number of veterinary visits by his good stockmanship.

Once the mare comes into season, a cervical swab should be taken. This cannot be done before as the cervix is only open when the mare comes into season. The swab result can be obtained within 24–48 hours so it is possible for the mare to be covered on this heat without wasting any more time.

Signs of oestrus

When a mare that is not in season is introduced to a stallion she may become violent, put her ears back, kick or even bite and the vulva will be tight and dry. She may also behave in this manner when pregnant. If she is in season, she will show at least one or more of the following signs: she will lean towards the teaser against the teasing board, hollow her back, straddle her hind legs, relax and raise her tail, evert the lips of the vulva so exposing the clitoris (commonly known as 'winking') and she will 'stand' for the stallion at covering. In some cases she may call to the stallion and if in the paddock she will walk the fence lines; she may even show these signs to another mare in the paddock. A 'silent' mare may only relax her tail or just stand quietly. It is up to the stud groom to recognise when to 'cover' the mare with the stallion, based on his observations of her behaviour. Some mares need teasing daily as they may only show for a very short time.

The teaser

It is usual with valuable stallions and mares to keep a stallion as a
teaser especially for the job. For safety reasons the teaser should
be of a suitable type and temperament. A pony stallion is often
used, preferably one of the British native breeds as they are easier
to manage and normally have easier temperaments. An older
thoroughbred or Arab can also be used. The temperament of the
stallion is extremely important as some are overactive and some
are too quiet. They need to show sufficient interest to 'chat up'
mares without frightening them, especially when dealing with a
maiden mare which has never been covered. Some mares 'show'
just from hearing the sound of horses' hooves; some need long
spells of teasing in the presence of the teaser before they will show.
It has even been known for the stud groom to have a tape record-
ing of stallion noises which he plays to the mares in the field —
mares which show easily react sufficiently to this method. A gelding
may be used but this is not always successful as some mares only
show to stallions. Using a vasectomised stallion may be a useful
and successful compromise.

Methods of teasing

The most common and safe method is to have a teasing board
placed in a convenient position near to the stables and paddocks,
in an enclosed area (Fig. 3.1). Alternatively the teasing board may
be in the fence line of each paddock used for mares and mares and
foals (Fig. 3.2). It is also possible to have a travelling board which
may be hung on the fence. 'Randy' mares will come up to the
stallion when he is led or ridden up to the paddock. Shy mares will
need to be caught up and taken to the teasing yard near the boxes
because the handler will have problems with a shy mare at teasing
should all the other mares be milling around the board to get
to the teaser. Some studs may exercise the stallions around the
property so that a constant eye may be kept on the mares' behaviour.
It is possible that when a mare is teased in the morning after a wet
or cold night she will not show to the stallion but as the day warms
up she may show later in the afternoon.

Stallions can be left loose in the stable and the mares brought to
the door, but this can be dangerous as the stallion could jump out.

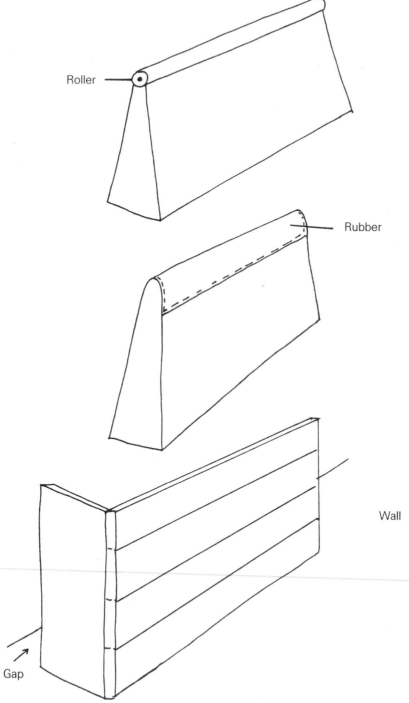

Fig. 3.1 Types of teasing board.

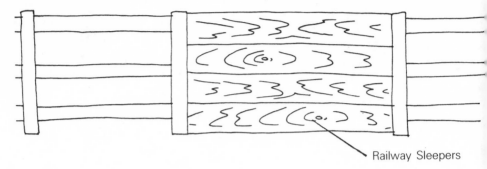

Railway Sleepers

Fig. 3.2 Teasing board in the fence line of paddock used for mares, or mares and foals.

It could also affect his temperament, making him more difficult to handle.

Some studs take a stallion into the field with the mares or let a smaller stallion run with the mares on a lunge to tease each mare at his own will. Both these methods involve greater risk to both handler and horse as does teasing mares over gates (Fig. 3.3) or over a rug hung on the fence (Fig. 3.4). In larger public studs dealing with many mares, staff safety is a prime consideration and it is necessary to work to a strict daily teasing routine. An example of one follows although this may not suit every stud.

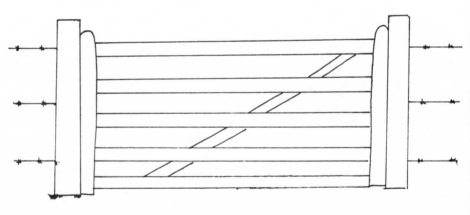

Fig. 3.3 Five-bar gate: a risky choice for teasing.

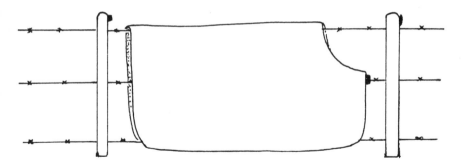

Fig. 3.4 Rug over fence: another risky choice for teasing.

Preparation for teasing

Before teasing or covering routines start the stallion or stallions should be racked up or chained to the wall of the stable. This is a safety precaution and is good stallion discipline. It eliminates any problems catching the stallion when he is required by his handler either to tease or cover a mare; if a stallion is within earshot of mares being teased or covered he may become upset, churn up his bed, become undisciplined and difficult to handle. It is preferable to have stallions stabled away from the teasing and covering yard.

When dealing with the competition stallion which works and competes as well as covering mares, he may well be rugged up. On hot days his rugs should be removed when he is racked up as the routine sparks off anticipation and the stallion may sweat.

The stud manager or groom should compile a daily list of mares required for teasing which is taken from the covering and teasing charts. Good communication is vital when running a busy stud and therefore staff must always be kept well informed to ensure the smooth running of the business. The mares can then be caught up if not stabled and either led straight to the teasing board or stabled until required. For convenience it is easier to leave the headcollars on. When handling mares and mares and foals it is not advisable to lead them directly in front of the stallion boxes during the covering season.

Fig. 3.5 The mare handler is suitably dressed and standing so that the mare will not hurt her should she strike out.

Handling of the mare during teasing

In most cases it is easier and safer to use a simple snaffle bridle when handling mares at teasing. However this may not be necessary with very quiet or maiden mares as one would not want to ruin a young mare's mouth. Instead she could be fitted with a chain or lunge line to the headcollar over the nose. In public or educational studs the handler should be safely and suitably dressed; good stout shoes, gloves and a hard hat should be worn when handling mares or stallions. Jewellery and earrings are both unsightly and dangerous as they can easily get caught up and cause injury to the handler.

If teasing is to take place in a teasing yard then the mare should be led and be made to stand quietly behind the board until the stallion approaches. She should be pushed close to the board so that the teaser is able to reach her with his nose. The handler should stand to the side of the mare, not in front of her so that should she strike out with her front feet she will not hurt the handler (see Fig. 3.5). As the teaser approaches, the handler should be aware and alert and ready for any violent movements from both stallion and mare. Once the mare has been tried and when instructed by the stallion handler, the mare handler should remove the mare from the teasing area back to the stable or paddock.

Handling of the stallion during teasing

It is safer and quicker to place the stallion bridle over the headcollar while the stallion is still racked up in his stable. The stallion's chain should then be attached directly to the bit before the stallion is untied. In some cases the stallion's chain and lead rein may be attached directly to the headcollar and passed over the nose of the horse for extra control. However quiet the stallion may appear a whip should always be carried as stallions, especially during the covering season, can become unpredictable and dangerous.

The stallion should be led quietly and firmly to the teasing board. He should never be allowed to rush, especially if the mare is a maiden as this could frighten her and she may not show any signs of oestrus. Let him first sniff her nose and then the body to the tail. Do not let him jump over the board (Figs. 3.6, 3.7, 3.8). Watch for the characteristic Flehman's posture − a curling of the lip when the stallion sniffs the mare which may indicate that she is in season (Fig. 3.9). Teasing boards should be so designed either well padded with rubber or with a roller fixed on the top to eliminate any injury.

Fig. 3.6 The stallion handler is letting the stallion quietly tease the mare.

Fig. 3.7 This mare is not in season. Note that this stallion wears a lunge rein as an overhead check.

When teasing is complete, to avoid the presentation of the stallion's heels to the mare, he should be trained so that he is pushed away from the mare and returned to the stables and tied up until required.

Covering

Covering is the term used for the mare and stallion mating.

Handling of the mare during service

The handler should always wear a hard hat, gloves and good boots or shoes when holding mares during covering. The mare should wear similar tack to that used for teasing. For reasons of hygiene, a clean tail bandage should be put on the mare and the vulva and area surrounding her genitals should be washed off with plain

Fig. 3.8 Mares may be encouraged to show signs of oestrus by rubbing their hindquarters with the whip handle.

warm water. Using disinfectant would kill all bacteria, both beneficial and harmful, thus lowering her resistance to infection.

Once prepared in this way the mare should be led to the covering yard where the thick felt covering boots should be put on (Fig. 3.10). If the mare has a foal at foot the foal should be left in the stable with competent handlers for safety reasons.

Young mares may object to the covering boots. However, having already mentioned the value of horses, it is essential that the stallion is not kicked by the mare. A twitch may also be applied to the mare to help control her (Fig. 3.11).

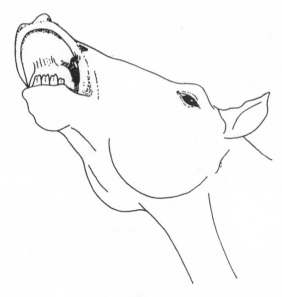

Fig. 3.9 Flehman's posture.

On the Continent, and sometimes in the UK, hobbles are used on the mares as a safety precaution but these also may be dangerous if a mare panics. Mares when fully in season stand quietly to a well mannered stallion without such restraints.

During service the mare's body must be kept straight but she may be allowed to walk slowly forward. If she is allowed to wriggle or step backwards this could cause damage both to herself and to the stallion.

Once the stallion has covered the mare successfully, the handler of the mare should pull her head towards himself. This will swing the hind quarters away from the stallion as he dismounts. After covering some mares are sensitive and may kick out, but if this method is adopted the danger of this occurring should be reduced. The kicking boots should be quickly undone and the mare walked slowly forwards to remove them and walked quietly for several minutes. This will reduce any straining and help to prevent any loss of valuable sperm. Before returning her to the paddock the tail bandage and bridle should be removed. It is important when releasing mares into a paddock to do so from a headcollar and not a

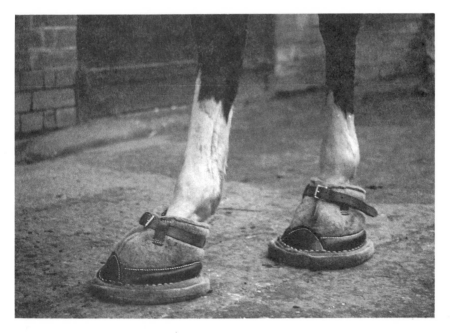

Fig. 3.10 Covering boots.

Fig. 3.11 A mare ready for covering.

bridle because the bit can easily become hooked on to the lower jaw and if the leatherwork is strong the mare's jaw can be broken. Some people believe that the foal should not be allowed to suckle immediately after its dam has been covered because the tickling sensation caused by the foal may aid the expulsion of the sperm. However, the stallion will ejaculate most sperm directly into the mare's uterus because the cervix is open and the possibility of some sperm loss is therefore unimportant. There are millions of live sperm in each ejaculate and only one is required to fertilise the egg!

Handling the stallion during covering

Handling of stallions during the stud season can be dangerous to both the horses and handlers. Therefore this job should be left to an experienced and confident stud person.

The stallion should be tacked up as for teasing and led quietly to the side of the mare. He must not be allowed to rush and jump straight onto the mare from behind; this could not only frighten the mare but could also cause her to kick (Fig. 3.12). He should not be allowed to mount the mare until he is completely drawn, i.e. the penis is completely erect. Once he is ready he is allowed to jump on the mare. Most experienced stallions will enter the mare unaided however, a younger stallion may need guidance by his handler (Fig. 3.13). In some cases during service the stallion may also need to have his front legs held by one or two handlers (Fig. 3.14). Mares in good physical condition are not so easy to cover because they are round and the stallion may not be able to get sufficient grip on the mare and fall off. If this happens serious damage to the stallion could occur which may even include a broken back. Signs of ejaculation or 'flagging' should be checked by watching the stallion's tail which will wave up and down at time of ejaculation (Fig. 3.15). Once the stallion retracts from the mare the penis should return to its original flaccid state almost immediately if he has ejaculated. The stallion's penis should be rinsed immediately in warm water to remove any mucus, and he should returned to the stable and racked up until all teasing and trying activities are complete. The stud groom should then complete any records necessary.

Fig. 3.12 A stallion is led quietly to the side of the mare.

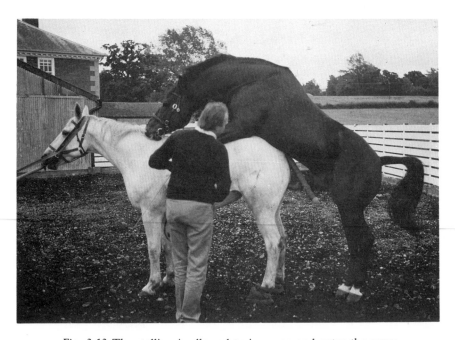

Fig. 3.13 The stallion is allowed to jump on and enter the mare.

Fig. 3.14 The stallion has his front legs held to help him stay on the mare.

This routine may vary considerably at different studs, but whatever the stud, teasing and covering should become a strict routine to ensure maximum safety to both horses and handlers.

After covering, the mare should be left and retried by the stallion 21 days after the first day she was first seen in season until 21 days from the last service date. Should the mare not return or come back into season she may be in foal; some six weeks later she should be tried again and if not in season she can be tested by the vet to see if she is in foal. With the development of ultrasound scanning, pregnancy results may be achieved a lot sooner, even as early as 15 to 16 days from the mare's last service date.

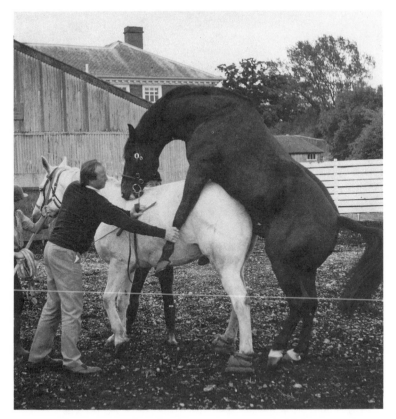

Fig. 3.15 The stallion's tail is 'flagging' indicating that ejaculation has occurred.

Pregnancy diagnosis

There are several methods of diagnosing whether or not a mare is in foal. The most modern and reliable is the ultrasound scanner. Alternatively pregnancy can be diagnosed by palpation of the uterus by the vet, or by taking urine and blood samples (though these last two methods are less reliable).

Scanning

The scanning machine is a very expensive piece of equipment and it is quite possible that only veterinary practices who deal with larger stud farms may be able to afford one! Mares can be scanned from as early as 15 to 16 days but more commonly from 18 to 21 days; the vet inserts the probe into the mare's rectum and the probe emits high frequency sound waves which are translated into a picture shown on the screen of the scanner. The conceptus will appear as a black spherical object within the uterus (Fig. 3.16). The principle is similar to that of pregnancy diagnosis in humans.

The advantage of this method is that much earlier results can be achieved but it must be noted that a mare can lose or reabsorb the conceptus or foetus at any time after conception. The scanner will also show whether or not the mare is carrying twins; if this is the case, the mare will nearly always lose one or both during her pregnancy. If she is left to go full term, it is very rare that both foals will live and develop into normal sized healthy horses. If twins are detected situated well apart, one may be 'squashed out' and to do this the vet will have to pinch the conceptus through the

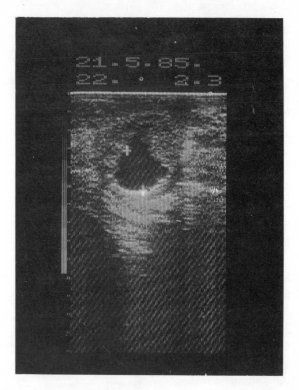

Fig. 3.16 A scanner photograph showing a 22 day old conceptus.

wall of the uterus by inserting his hand into the rectum. It would be advisable for the mare to be rescanned a few days later to make sure there is only one surviving pregnancy. If the twins are very close together in the uterus, it will be much more difficult to abort one of them and if this problem should occur early in the stud season, it may be advisable to inject the mare with prostaglandin to abort her and start again.

Palpation of the uterus

This is generally done from 42 days onwards but an experienced vet will be able to tell earlier than this. He feels the uterus through the wall of the rectum and if there is a swelling in one of the uterine horns it is assumed that the mare is pregnant. He may also feel the 'tone' of the uterus; good tone − i.e. a tight or firm feel − is a good sign of a pregnant and healthy mare. Bad tone − i.e. a flaccid, large uterus − is a sign of a mare not in foal or a mare that has just foaled and not yet returned to her normal physical state. It may be a sign of a mare not able to conceive at all.

Blood testing for pregnancy

A blood sample is taken approximately 45 to 90 days from the last service day and tested for levels of pregnant mare's serum gonadotrophin (PMSG). However, this test can give false positive results as mares can have PMSG in their system and still not be in foal. This occurs if the foetus is reabsorbed after day 45 of pregnancy and the endometrial cups which produce PMSG have been formed.

Urine sample

Urine is taken from approximately 100 days after the last service and tested for levels of oestrogen which rise dramatically during pregnancy. This method, however, is not practical, mainly because of the difficulty in collecting urine samples and the delay between covering and obtaining the results.

Mare management

An infoal mare should be treated as closely to nature as possible during pregnancy. Overfeeding and too much fussing is detrimental to the mare's health and subsequent condition of the newborn foal. From the time she is tested in foal until Christmas she should only have good quality pasture. Her body condition should be observed and she should only need to be fed just before she starts to lose condition. She may be brought in at nights although exercise is most important to keep her physically fit and to ensure an easy foaling. She should be wormed regularly at four to six week intervals until she foals and her feet trimmed every six to eight weeks. She should also have been kept up to date with influenza and anti-tetanus injections.

Successful stallion and broodmare management depends on correct year-round care; management should ensure that horses are kept in as natural a way as possible so that they are happy, relaxed and healthy.

4 The reproductive system of the mare

The genital tract

The breeding organs consist of two ovaries, the uterus, cervix and vagina, which are suspended within the abdominal cavity by sheets of strong connective tissue (Figs. 4.1 and 4.2). Knowledge of how these organs work is important in understanding the sexual functions of the mare.

The ovaries

The *ovaries* of the mare are situated between the last rib and the point of the hip; they are roughly bean-shaped and measure about 7 x 4 cm (Fig. 4.3). Each ovary contains many thousand female gametes called eggs or *ova*; every mare is born with a full complement of eggs and does not produce more during her lifetime. Each egg is stored in a sac of fluid called a *follicle* which is surrounded by a fibrous mass or stroma. Prior to ovulation the follicle enlarges up to a size of 4 cm and is known as the *Graafian follicle*. The mature or 'ripe' follicle bursts at the ovulation fossa situated at the pinched-in face of the ovary, and is released into the Fallopian tubes.

The Fallopian tubes

The *Fallopian tubes* or oviducts are two coiled tubes 20–30 cm long which carry the ovulated egg from the ovary to the uterine horn. After ovulation the egg is guided into the Fallopian tube by the fringed end of the tube which partially overlaps the ovulation fossa. There the sperm meet the egg and if fertilisation occurs the resulting embryo is moved along the tube and into the uterus.

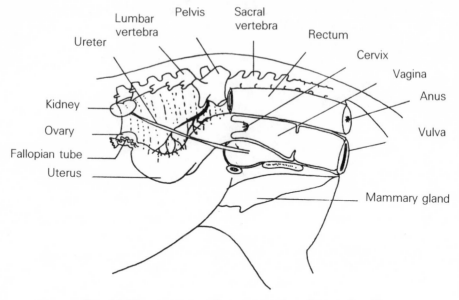

Fig. 4.1 The mare's breeding organs viewed from the side.

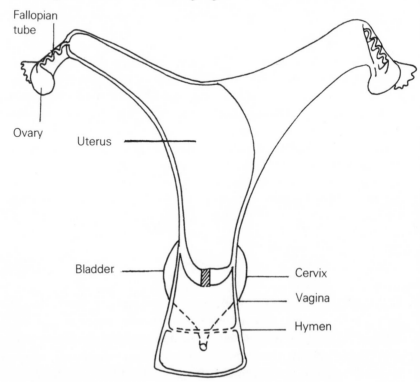

Fig. 4.2 The mare's reproductive organs viewed from above.

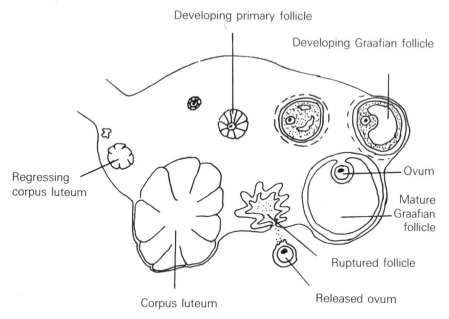

Developing primary follicle

Developing Graafian follicle

Regressing
corpus luteum

Ovum

Mature
Graafian
follicle

Ruptured follicle

Corpus luteum

Released ovum

Fig. 4.3 Events occurring in the ovaries during the oestrous cycle.

The uterus

The *uterus* is a hollow muscular organ in which the foetus develops. It is 'Y' shaped, comprising two horns, a body and a neck (cervix) and is suspended in the abdomen by the *broad ligaments*. The horns are about 25 cm long, and the body 18−20 cm in length and 10 cm in diameter. The body of the uterus connects to the outside via the cervix, vagina and vulva.

The cervix, vagina and vulva

The uterus and vagina are separated by a muscular constriction or neck called the cervix. The cervix protects the mare and foetus from uterine infection during pregnancy and when the mare is not in season by being tightly closed. However, when the mare is in season or giving birth the cervix can relax and dilate allowing the entry of semen or the expulsion of the foal.

The vagina is a passageway between the cervix and the external opening of the mare's genital tract, the *vulva*. The vagina protects

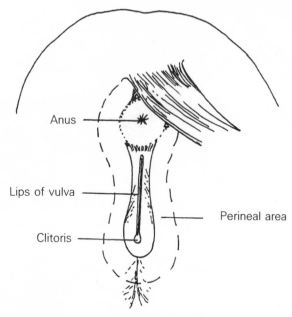

Fig. 4.4 The external genitalia of the mare.

the uterus from contamination from the outside and is frequently subject to infection. In the maiden mare the vagina and vulva are usually separated by a thin fold of tissue or *hymen*. The urethra carries urine from the bladder and enters the vagina so that urine is expelled through the vulva.

The vulva acts as a protective door just below the anus (Fig. 4.4). The lips of the vulva are arranged vertically on either side of the vulval opening. Between the vulval lips at the lower end of the vulva lies the *clitoris* which protrudes between the vulval lips when the mare is in season, an action called *winking*.

The oestrous cycle

The mare is said to be *seasonally polyoestrous*, in other words she has a breeding season during which she has several oestrous cycles or periods (Fig. 4.5) when she will accept the stallion. Normally the mare shows sexual activity during late spring, summer and early autumn and not during the winter. This is an attempt to ensure that the foal is not born too early or too late in the year which would reduce its chances of survival. However, man has intervened and decreed that all registered thoroughbreds should have their birthday on 1 January. This has led to breeders trying to get mares to foal earlier in the year than would normally occur.

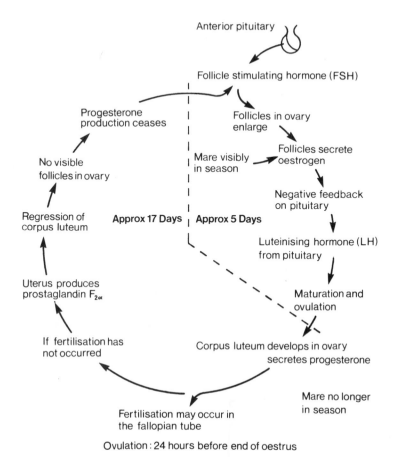

Anterior pituitary

Follicle stimulating hormone (FSH)

Progesterone
production ceases

Follicles in ovary
enlarge

No visible
follicles in ovary

Mare visibly
in season

Follicles secrete
oestrogen

Regression of
corpus luteum

Approx 17 Days

Approx 5 Days

Negative feedback
on pituitary

Luteinising hormone (LH)
from pituitary

Uterus produces
prostaglandin F$_{2\alpha}$

Maturation and
ovulation

If fertilisation has
not occurred

Corpus luteum develops in ovary
secretes progesterone

Mare no longer
in season

Fertilisation may occur in
the fallopian tube

Ovulation : 24 hours before end of oestrus

Fig. 4.5 Percentage of mares ovulating during the year. Note that the maximum fertility lies outside the thoroughbred breeding season.

Thus the stallion owner and vet have needed an understanding of the oestrous cycle in order to get mares in foal earlier in the season and this importance has been emphasised with the introduction of artificial insemination and embryo transfer.

The pituitary gland and hormones

The physical behavioural changes that occur during the oestrous cycle of the mare are controlled by chemical substances called *hormones*. Hormones are produced by glands and released into the bloodstream so that they travel around the body and exert specific effects on *target organs*. The 'master gland' of the body is the *pituitary gland*. This is a small body situated immediately under the base of the brain (*hypothalamus*) and is divided into anterior and posterior parts. The anterior pituitary produces two *gonadotrophins* (hormones which affect the reproductive organs) called follicle stimulating hormone (FSH) and luteinising hormone (LH) which act on the ovaries.

The oestrous cycle has two components: oestrus, 5–7 days during which the mare is in season and is receptive to the stallion, and *dioestrus*, 14–16 days during which she rejects the stallion.

Oestrus

FSH from the pituitary stimulates the growth and development of a small number of follicles within the ovary. They increase in size from microscopic proportions to reach a diameter of several centimetres. One, the Graafian follicle, will outgrow the others and when approximately 4 cm in diameter it will rupture or ovulate (see Fig. 4.3). The developing follicles act as glands and secrete the hormone *oestrogen*. As the levels of hormone rise a gradual change will be seen in the behaviour of the mare; instead of rejecting the teaser violently she will alternate between accepting his advances and rejecting them, and she is said to be 'coming on'.

Eventually she will come into season and allow herself to be covered by the stallion. Oestrogen also causes changes in the reproductive tract; the vulva and cervix relax allowing the stallion to ejaculate directly into the uterus, the blood supply to the uterus increases and vaginal secretions increase.

About 24 hours before the end of oestrus ovulation occurs, stimulated by LH from the pituitary; the egg is released through the ovulation fossa and passes into the Fallopian tube. If the egg is fertilised it travels along the tube reaching the uterus five days after fertilisation. If it is not fertilised the egg remains within the tube and disintegrates.

Dioestrus

Once the follicle has ruptured and released the egg a blood clot forms within the empty cavity. Special granulosa cells invade the blood clot transforming it into a mass of yellow tissue called the *corpus luteum* or *yellow body*. The corpus luteum acts as a gland and produces the hormone *progesterone*. Progesterone reverses the effect of oestrogen; the mare goes out of season (generally 24 hours after ovulation), the cervix becomes tight and pale and the uterus becomes firm and less secretory. Progesterone effectively prepares the uterus for receiving the fertilised egg and is essential for maintaining the pregnant state.

If fertilisation takes place the uterus recognises the pregnancy and the corpus luteum remains functional and supports the pregnancy by maintaining adequate levels of progesterone. However, the mare is unable to come back into season if there are appreciable levels of progesterone in the blood; if the egg is not fertilised it is important that progesterone secretion is halted. In order to accomplish this the uterus produces a substance called *prostaglandin* $F_{2\alpha}$ which 'kills off' the corpus luteum, progesterone secretion stops and the mare is allowed to come back into season.

Summary

The pituitary gland produces FSH which stimulates the growth of follicles in the ovary. The ovaries produce oestrogen which brings the mare into season. High levels of oestrogen cause the pituitary to reduce the secretion of FSH and increase output of LH. Ovulation occurs and a corpus luteum forms which produces progesterone. Falling oestrogen and rising progesterone levels cause the mare to go out of season. If fertilisation does not take place the uterus releases prostaglandin which causes the corpus luteum to regress and progesterone levels to fall. This stimulates the anterior pituitary to secrete FSH and restart the cycle.

This is a simplified account of an ideal cycle; in practical terms the ideal rarely happens.

5 The reproductive system of the stallion

The genital tract

The breeding organs of the stallion consist of two testes, epididymis, vas deferens, the accessory sex glands and the penis (Fig. 5.1). If the stallion handler is to identify potential fertility or behavioural problems a basic understanding of the stallion's reproductive organs and normal sexual behaviour is very important.

The testes

The *testes* are two oval structures contained within folds of skin (scrotum) between the thighs of the stallion. Each is about 12 cm long and weighs about 300 g (Fig. 5.2).

During foetal development the testes are located close to the kidneys but prior to birth they migrate and pass through an opening in the abdominal wall (inguinal ring) and into the scrotum. In some cases one or both testes fail to enter the scrotum and remain within the abdomen. This horse is known as a *cryptorchid* or *rig*.

Each testis consists of soft reddish-grey material and is covered by a strong fibrous membrane. Sheets of this membrane divide the testis into lobules containing the small coiling *seminiferous tubules*. Within these small tubes are the cells which divide to produce spermatozoa, and those cells responsible for the production of the male hormone *testosterone*. Testosterone is responsible for 'stallion-like' behaviour and the secondary male characteristics, e.g. crested neck. The seminiferous tubules join up and converge towards the front of the testis and pass through the covering coat into the epididymis.

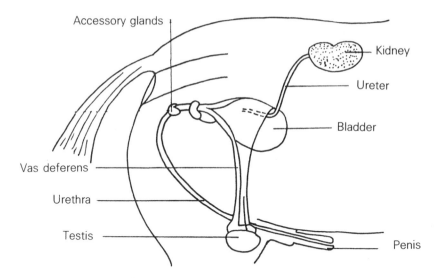

Fig. 5.1 The reproductive organs of the stallion.

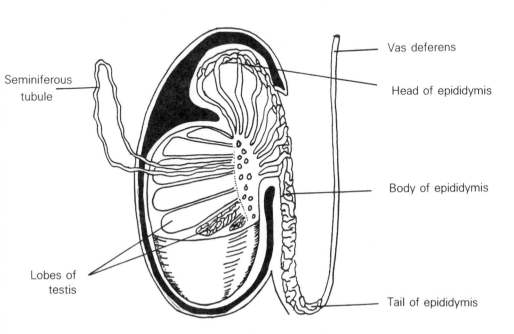

Fig. 5.2 Cross-section of the testis.

The epididymis

The epididymis is a U-shaped structure attached to the top of each testis, the function of which is to concentrate and store sperm. The epididymis consists of a head, a body and a tail. The head is composed of several tubes which unite to form a single *epididymal duct* in the body of the epididymis. This duct is very long and convoluted and passes on into the tail of the epididymis and enters the vas deferens.

The vas deferens

The vas deferens is a muscular tube that carries sperm and the associated fluids from the epididymis to the urethra. It passes through the inguinal canal as part of the *spermatic cord* and then separates from the other spermatic vessels and runs backwards to enter the pelvic cavity. As it nears the bladder the walls of the tube enlarge to form the *ampulla* after which it decreases in size to unite with the *urethra* and enter the *penis*. The urethra is a tube that carries urine from the bladder to the end of the penis.

The penis

The penis is used to deposit semen in the female reproductive tract; it is composed of erectile tissue and terminates at the *glans penis* which is the enlarged free end of the organ. Erection occurs when the penis becomes engorged with blood when the stallion is sexually stimulated. The base of the glans penis is surrounded by a prominent margin through which the end of the urethra projects. The free portion of the non-erect penis is covered by the sheath or *prepuce* which is a double fold or pocketing of skin. Large glands in this area produce a fatty cheese-like substance called *smegma*. If the smegma is not removed regularly it can become foul-smelling and a source of irritation and infection to the horse.

The accessory glands

During ejaculation the accessory glands secrete about 60–90 per cent of the total seminal fluid volume; they consist of the seminal vesicles, the prostate and the bulbo-urethral glands.

(a) *The seminal vesicles*: These are two elongated sacs that lie either side of the bladder. Smooth muscle contractions during ejaculation empty the secretions into the urethra.

(b) *The prostate gland*: This is also situated at the beginning of the urethra and it releases its milky secretion into the urethra through about fifteen ducts.

(c) *The bulbo-urethral glands*: These two glands lie either side of the urethra and release a clear fluid that flushes the urethra prior to ejaculation.

Semen characteristics

Semen consists of sperm and seminal fluid from the accessory glands. Semen quality is a valuable indicator of stallion fertility and it is important to be able to recognise abnormal characteristics.

The spermatozoon

The spermatozoon or sperm is the male sex cell and it carries the stallion's genetic contribution to his offspring. It consists of a head, a midpiece and a tail (Fig. 5.3); the head consists mainly of a nucleus which carries the genetic material or chromosomes. As the sperm are formed in the testis the number of chromosomes is reduced by half (Fig. 5.4) thus it is said to carry the *haploid* number or 32 chromosomes. The egg also contains the haploid number and when fertilisation takes place and the two cells fuse the resulting fertilised egg has 64 chromosomes. This is the *diploid* number found in all body cells. The midpiece connects the head to the tail and contains the energy stores which motivate the tail. The tail is long and whip-like and allows the sperm to be highly motile and capable of swimming great distances. The average ejaculate of 40 to 120 ml contains about 100 to 150 million sperm per cubic millimetre.

Seminal plasma

Seminal plasma plus sperm is called *semen*: it is a milky white gelatinous liquid, and is secreted by the accessory glands, the vas deferens and the epididymis. Seminal plasma serves as a transport

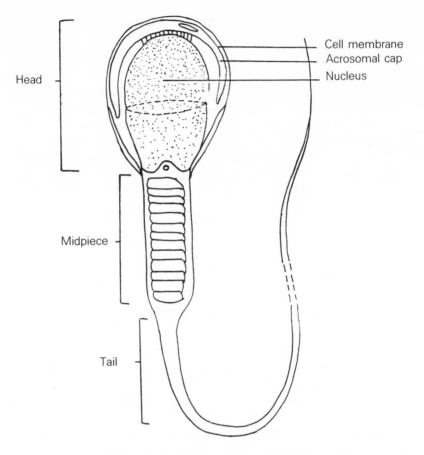

Fig. 5.3 A normal mature sperm cell.

medium for the sperm and contains energy sources and chemicals capable of maintaining sperm as they mature within the female tract. These chemicals should be present in characteristic amounts in the semen as shown in Table 5.1.

Table 5.1 Properties of semen.

Volume of semen	30–250 ml
Number of sperm	30–600 million per cubic millimetre
Abnormal sperm	less than 35%
pH	average 7.3
Specific gravity	1.012
Ergothioneine	7.6 mg/100 ml
Citric acid	26 mg/100 ml
Fructose	15 mg/100 ml
Phosphorus	17 mg/100 ml
Lactic acid	12 mg/100 ml
Urea	3 mg/100 ml

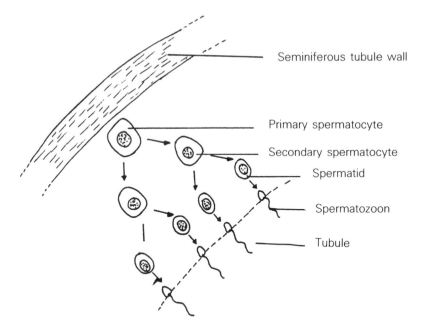

Seminiferous tubule wall

Primary spermatocyte

Secondary spermatocyte

Spermatid

Spermatozoon

Tubule

Fig. 5.4 Sperm production.

These characteristics have a wide range of normal values with season of the year having a marked effect. For example, citric acid, ergothioneine and sperm numbers decrease during the winter and increase during summer. The concentration of sperm may increase by as much as seven times.

Semen can be divided into four parts:

(a) *Pre-sperm fraction*: A clear fluid that cleans and lubricates the urethra prior to ejaculation.
(b) *Sperm-rich fraction*: About 30–75 ml of fluid containing 80–90 per cent of the total number of sperm. Has a high concentration of *ergothioneine*.
(c) *Post-sperm fraction*: A sticky gel secreted by the seminal vesicles after ejaculation, very variable in volume.
(d) *Tail-end fraction*: Contains very little gel and few sperm.

Hormonal control

As with the mare, the production of sex cells and sexual activity is controlled by the pituitary gland. FSH stimulates the growth and formation of sperm in the testes, while LH causes the *interstitial*

cells of the testes to release the male hormone *testosterone*. Testosterone promotes the male characteristics such as crest development
and sexual drive.

Semen evaluation

A thorough evaluation of a stallion's semen aids in the diagnosis of
subfertility and infertility by helping to show how well the organs
that produce the semen are working. However, the job of stallion
'fertility testing' is difficult due to the wide variation in the values
of constituents of 'normal' semen. A methodical approach is necessary, gathering as much information as possible to build up an
overall picture. This means that it is vital to study the stallion's
breeding history, to give him a physical examination, to perform a
bacterial examination and to observe his sexual behaviour as well
as to evaluate a semen sample. In practice, this technique is adopted
if stallions are used for artificial insemination or are failing to get
mares in foal.

Collection of semen

The most efficient way to collect semen is using an artificial vagina.
This technique will be discussed in greater detail in the chapter
dealing with artificial insemination.

Semen examination

Initially the sample will be observed for appearance and volume,
then the motility and longevity of sperm are measured.

(a) Appearance: Milky-white, no clots, no unusual smell.
(b) Volume: Total volume should be 60–120 ml and gel-free volume
    about 50–100 ml. However, as previously stated, these figures
    are very variable, depending on the stallion and the time of
    year.
(c) Motility: This is a very important measure of stallion fertility.
    Sperm should move forward rapidly; those that move in tight
    circles are unlikely to fertilise an egg and are not considered to

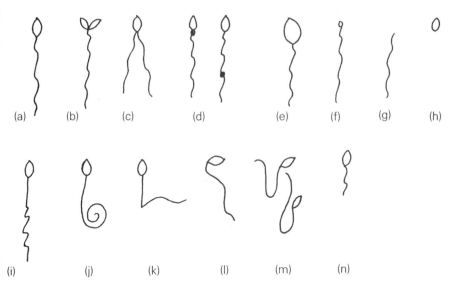

Fig. 5.5 Sperm abnormalities (a) Normal; (b) Twin head; (c) Twin tail; (d) Immature; (e) Giant head; (f) Small head; (g) Headless; (h) Tailless; (i) Crooked tail; (j) Coiled; (k) Bent tail; (l) Bent; (m) Shoehook; (n) Short tail.

be motile. Sperm from a normal fertile stallion should have an immediate motility of 60 per cent.

(d) *Morphology*: This is the actual appearance of the sperm (Fig. 5.5). The sample should contain at least 65 per cent normal sperm. Abnormal sperm indicate a malfunction in the sperm production process. Primary abnormalities, e.g. having two heads or two tails, suggest a failure of spermatogenesis in the testes; secondary abnormalities suggest a failure of the sperm to mature in the epididymis; and tertiary abnormalities are due to damage during or after ejaculation. A high proportion of primary abnormalities can be serious but secondary abnormalities may only be due to over-use of the stallion, the frequent ejaculations not giving sperm enough time to mature in the epididymis.

(e) *Live-dead ratio*: This ratio is highly variable but at least 50 per cent should be live.

(f) *Sperm concentration*: The normal range for sperm concentration is 30−600 million sperm per ml of semen and a total of 500 million sperm are needed to obtain optimal conception rates. Most stallions are well in excess of this minimal requirement.

(g) *Acidity and alkalinity*: A pH of between 7.3 and 7.7 is considered

normal. Semen that is abnormally alkaline may have been infected with urine or may indicate a malfunction of the accessory glands.

(h) Longevity: At room temperature sperm should stay alive and motile for 8—24 hrs.

Table 5.2 Normal semen parameters.

Red blood cells	Less than 500/mm^3.
Morphology	At least 65% normal.
Live/dead count	At least 65% live cells.
Motility	At least 40% actively motile.
Longevity	At least 40—50% alive after 3 hrs at room temperature; at least 10% alive after 8 hrs (with 2—5% motility).

6 Foaling, foaling problems and foal diseases

Foaling

The preparation and organisation of facilities and equipment prior to foaling is vital to ensure safe foaling.

The foaling box

A large draught-free foaling box or stable is required, the bigger the better but approximately 15′ × 15′ (4.5 × 4.5 m) is adequate for a 16 hh plus hunter broodmare to foal in. The foaling box or boxes will vary from stud to stud but must be free from any sharp protruding objects. Ideally, there should be no corners, as a mare gets up and down during foaling quite frequently; cut off corners eliminate the risk of the mare getting cast or stuck in the corner whilst foaling. There should be no feed bucket or water buckets on the floor because, should the mare foal without warning, a newly born foal may easily get stuck or be drowned in a bucket on the floor. Automatic water and a high manger will eliminate this risk. It is also far safer to put hay on the ground rather than in a hay net as a foal could easily become caught in the net when playing.

A good thick bed of straw well banked at the edges is necessary. The banks will help prevent the mare getting cast, as quite often a mare will go down and brace herself against the wall whilst foaling. The box should be regularly mucked out prior to foaling and steam-cleaned, painted and thoroughly disinfected between foalings.

Closed circuit television

It is possible to have a closed circuit television installed, which will enable the stud groom to observed foalings from the comfort of his home or sitting-up room. Valuable mares should never be left unattended during this period; about 90 per cent of mares may foal unaided but of the 10 per cent which have problems approximately 90 per cent of these foals can be saved with human help. The closed circuit TV can be left running and is quiet; indeed, some people watch the whole foaling from this and only enter the stable when a problem arises.

Foaling alarms

A foaling alarm is an electrical device attached to a roller which is fitted to the broodmare when her foaling is imminent. Her temperature will rise just before foaling and this triggers off an alarm warning the attendant she is about to foal. This system in conjunction with the closed circuit TV is ideal, though rises in temperature through other reasons, e.g. hot weather, could easily set off the alarm.

Foaling equipment

Clean towels: For rubbing down the newly born foal and general use.
Foal rug or jersey: It is fashionable to produce early foals and should the weather be particularly inclement, a small rug for the foal may be required. However, an old polo-necked jersey is equally adequate.
Enema: To aid in the release of any meconium, a human enema acquired from any medical supplier, already prepared, is a safe and ideal treatment for this job.
Liquid paraffin (or similar): Should meconium retention not be cleared by the use of an enema, then an oral dose of liquid paraffin may be required.
String: Once the mare has foaled, she may not 'cleanse' directly and the trailing afterbirth should be tied up until she does so.
Water and soap: For washing hands, etc.
Rug for mare: Needed for early foalings. Maiden mares may suffer from shock after foaling.
Cotton wool: For general veterinary uses.
Antiseptic spray, iodine based solution or wound powder: Needed

for dressing the foal's umbilical cord stump immediately after foaling.
Feeding bottle and teat: This is required should the mare die during
foaling and the foal need feeding, or if the mare is not producing
enough milk of her own and the foal requires supplementary feeding
or merely to ensure that the foal gets enough colostrum. Colostrum is
the mare's first milk and contains lots of antibodies which help the
newborn foal fight disease until it is able to produce its own
antibodies to combat any infections. In the long run, it is more
practical to teach the foal to drink any milk directly from a bucket.
Mare's milk replacer: This is dried milk sold in bags which can be
bought within hours from the manufacturers. No stud should
commence the foaling season without having milk replacer in stock.
There is no need to waste it as it can be fed along with concentrates
when the foal is weaned.
Frozen colostrum: Colostrum can be milked from an early foaling
good milking mare and kept frozen in a deep freezer. Should a
mare die during foaling, it is essential that the foal gets this for
reasons already stated. If colostrum is not available, then a course
of antibiotics given to the foal may give adequate protection.
Infra-red heat lamp: To warm mare and foal if required.
Headcollar: All mares should be foaled in a headcollar; maiden
mares may become difficult after foaling and wearing a headcollar
makes them easier to catch.
Tail bandage: The mare's tail should be bandaged as for covering;
there is nothing worse than a cold wet tail around the mare's hocks
after foaling.
The vet's telephone number: This should always be well posted in
case of an emergency.
Lubricant: Lubricant gel may be required by the stud groom or vet
during a difficult foaling.
Calving or foaling ropes: These should be at hand for use at a
difficult foaling but should only be used by the experienced stud
person or vet. The veterinary cupboard should always be kept well
stocked for emergencies.

Signs of foaling

A mare will foal approximately 340 days from her last service date,
but this can vary considerably and mares can be anything up to
seven weeks late and foal quite normally. In these cases veterinary

advice should be sought. If the mare foals after 325 days of pregnancy, this is as good as full-term and the foal will live. A foal is described as being premature if it is born between 300 and 325 days and the foal may even die. Foals born at less than 300 days of pregnancy will have no chance of survival.

These figures are only approximate, however, and the mare will show various signs, which the mare owner must be careful to look for, as the foaling become imminent. As she gets nearer to foaling she will get much heavier in weight and slower in her movements, she is less likely to exercise herself and her legs could fill. She may need walking in-hand daily as even when turned out in a paddock she may stand by the gate all day until she is brought in at night. About a month before she is due to foal her udder will start to swell but this will go down after exercise. During the last week or so before foaling, however, it will remain full and tight even after exercise. Some mares bag up rapidly and foal with little or no warning.

Eventually a wax-like substance will appear on the end of each teat; some mares may even run milk. Once the mare has 'waxed up' she may foal at any time and should not be left unattended. The muscles and ligaments round the pelvis will become relaxed and slack to enable the foal to pass through. This will show by the sagging of the muscle either side of the tail, the vulva also becoming long and inflamed.

Foaling

There are three stages of foaling:

The *first stage* (see Figs. 6.1 and 6.2) is from first contractions until the breaking of the first waters. This, as in humans, may be a very short time or several hours.

The *second stage* is from the breaking of waters until the birth of the foal. If the mare does not foal within approximately half an hour after the breaking of waters, the vet should be called.

The *third stage* is from the birth of the foal until the expulsion of the afterbirth. Mares normally foal late at night and, should the mare not have cleansed by morning, the vet will be required to assist her.

(a) First stage

The mare will become restless and pace around uneasily. She may get up and down and swish her tail and show other signs of pain

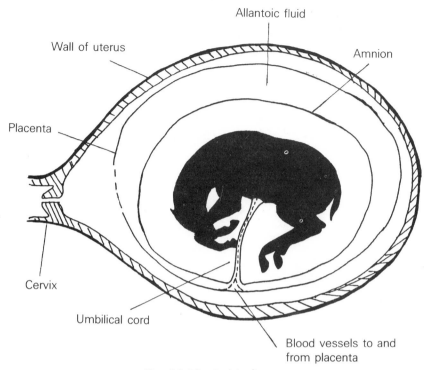

Fig. 6.1 The foal in the womb.

such as kicking and looking at her flanks. Depending on the mare, she may become hot, her temperature will certainly rise and she could sweat profusely. At this stage she should be wearing her headcollar and a tail bandage. Mares will normally foal lying down, but some will foal standing up.

Eventually, the first waters will break which must not be confused with the mare urinating. A brownish urine-like liquid will gush from the vulva because the foal's fore feet have punctured the weak spot of the placenta, thus releasing the outer waters or allantoic fluid.

(b) Second stage

If all is going well an opaque whitish blue bag will appear (the amnion). This has a balloon-like appearance and should contain the foal's first foot. At this stage the stud groom may insert his hand into the mare's vagina to check that the foal is presented correctly. The foal is presented in a diving position (see Fig. 6.2) and he should be able to feel the foal's second leg and head in the

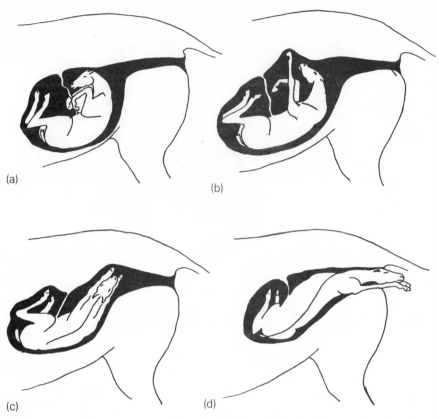

(a)

(b)

(c)

(d)

Fig. 6.2 During the first stage of labour the foetus gradually shifts from a position on its back, rotating until its head and forelimbs are extended in the birth canal in a diving position.

correct position. The second leg always follows the first in a manner which will slant the foal's shoulders, enabling it to pass more easily through the mare's pelvis. It is common for the mare to get up and go down again at this stage − this is 'nature's way' of correcting the foal's position should it be twisted in any way (Figs. 6.3 to 6.9).

After further contractions from the mare the foal's head will then be seen covered by the amnion. The amnion will often break naturally whilst foaling but occasionally may need assistance as a foal can easily suffocate inside the bag. It must be remembered that the foal is receiving vital oxygen from the mare's blood which is passing through the umbilical cord at this stage, and if the mare starts to cleanse or the cord breaks prematurely, the foal could die or suffer brain damage through lack of oxygen (dummy foal).

Fig. 6.3 The amnion containing one forefoot.

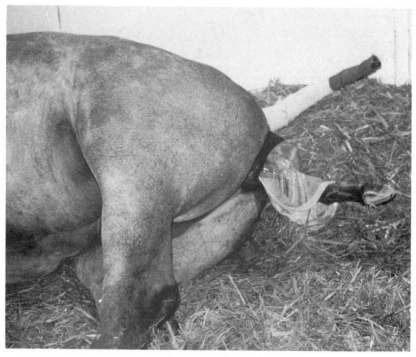

Fig. 6.4 The amnion has broken and the foal's head can now be seen.

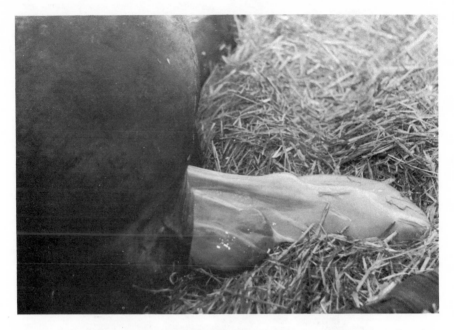

Fig. 6.5 The amnion has not broken naturally and care must be taken that the foal does not suffocate inside the bag.

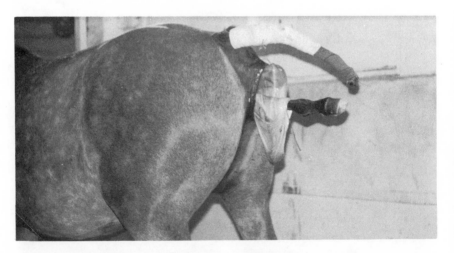

Fig. 6.6 The mare may get up and lie down again at this stage.

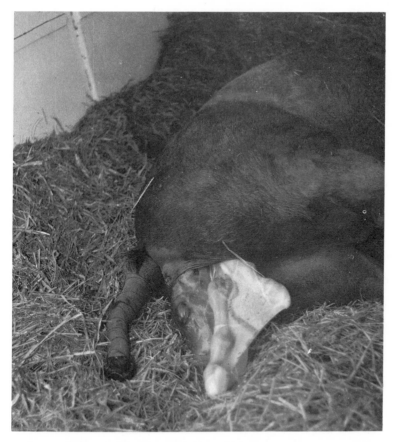

Fig. 6.7 The mare continues to push and the foal's head emerges.

Fig. 6.8 The foal's hindquarters may rest inside the mare's vulva for several minutes.

Fig. 6.9 Care must be taken not to break the umbilical cord too soon.

Once the shoulders are out the mare may rest and should not be disturbed. Never rush a normal foaling: it is so easy to panic and start pulling out a foal unnecessarily. However, if there is a problem, this should be recognised immediately and suitable action taken.

After some less violent contractions by the mare, the foal should be born; at this stage the umbilical cord should still be intact, and both mare and foal should be left still and quiet for as long as possible. The foal may even have left its hind legs resting inside the mare's vulva, which is the natural way to keep the mare down.

More often than not the cord will break naturally when either mare or foal moves but sometimes assistance is needed; gentle pressure to the foal's abdomen with the flat of the hand, the cord emerging between the fingers and a steady pull with the other hand will sometimes assist in the breaking of the cord. Once broken, it must be ensured that any bleeding stops. At this stage the umbilical stump should be sprayed with purple spray or some other antiseptic product; infection could rapidly enter this area and could cause joint ill (see foal diseases).

If the mare has not moved the foal should be brought round to the mare's head. It is important at this stage for mare and foal bonding to take place. The sex of the foal should be established and he or she should be towelled dry with a handful of straw, although the older, more experienced, mare will take this job into her own hands and lick her foal dry.

(c) Third stage

The mare may cleanse immediately, cleansing being the expulsion of both 'sacs' — the amnion and placenta — along with the umbilical cord. If she shows no signs of imminent cleansing, the trailing afterbirth should be doubled up and tied with a piece of string (Fig. 6.10). This creates a natural weight to assist with the cleansing; if it is not tied-up the mare could step back and tear away the placenta.

Fig. 6.10 The trailing afterbirth should be tied up to prevent the mare stepping on it.

Once expelled, the mare may actually eat the placenta. If not, the placenta should be examined to make sure that it is intact. Any piece of placenta which may have been torn off and left inside the uterus may cause infection and this could lead to septicaemia (blood poisoning), the results of which could be fatal (Figs. 6.15, 6.16).

The attendant should not leave the mare and foal alone until both are on their feet and the foal is drinking from the mare unaided (Figs. 6.11–6.14). A slow foal may need assistance to find the teat or the mare may have to be milked and the foal given a pint or two of milk from a bottle to ensure colostrum intake. Never twitch a mare immediately after foaling if she becomes difficult and refuses to let the foal drink. Twitching could cause internal haemorrhaging and be fatal. Gentle assistance may be needed to help a shy foal to drink but remember that if you force the foal too much, it can only make things worse. A foal will absorb antibodies from the mare's milk for as little as 12 to 48 hours, after which it will manufacture its own.

Table 6.1 provides a list of normal facts and figures − a useful reference to check that the foaling has been normal and that the normal foal is healthy.

Fig. 6.11 The mare will lick the foal and establish a bond between them.

Fig. 6.12 The foal should be quickly checked over for abnormalities and should look alert.

Fig. 6.13 The mare and foal should be left alone as much as possible to establish their bond.

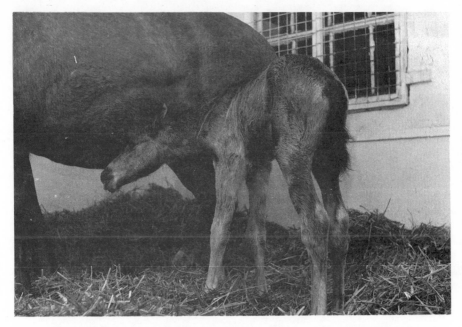

Fig. 6.14 The foal should not be left unattended until it has suckled.

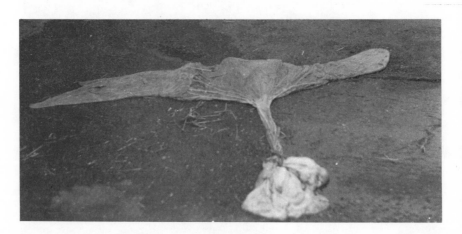

Fig. 6.15 The placenta must be checked.

Fig. 6.16 The hippomane, an object found in the afterbirth.

Table 6.1 Normal foal facts and figures.

	Average	Normal range
Length of gestation	340 days	(320−360)
Bagging up of mammary gland	10 days	(2−30)
Running milk	0 days	(0−10)
Wax on teats	3 days	(0−6)
Duration of parturition stages (Foaling)		
First stage	30 mins	(10 mins−2 days)
(steaming up − breaking water)	30 mins	(5−60 mins)
Second stage	5 mins	(5−15 mins)
(breaking water − hips through pelvis)	2 hrs	(20 mins−12 hrs)
(breaking water − showing amnion)		
Third stage		
(birth − passing foetal membranes)		
Veterinary attendance		Not more than eight hours

Foal birth (Completion of birth is when hips are free of mare's pelvis)

Birth to breathing	90 secs
Birth to breaking umbilical cord	5−15 mins
Birth to suck reflex	up to 20 mins
Birth to standing	20−100 mins
Birth to sucking mare	60−120 mins
Birth to passing meconium	0−2 days

Observations worth noting include:
Passing of first meconium. This will be observed by foal straining a lot.
Urination: time first noticed.
Body temperature (99°F−101°F)
Respiration rate   (30−40 per minute)
Heart rate   (80−140 per minute)

Mare's colostrum. Normally contains high levels of antibodies measured as globulins, but if she has run milk for any length of time the quantity falls.

	Normal colostrum	*Normal range*
Total serum protein	15.32 g/100 ml	(13−18)
Albumin	1.86 g/100 ml	(1.4−3)
Globulin	13.46 g/100 ml	(11−16)

It is very important to make sure the mare and foal are exercised the following day. If the weather is inclement, it is a good idea to lead out the mare in a paddock with the foal loose; this will also assist the foal to pass its meconium (see foal diseases). The following day it is advisable to have the foal injected with antibiotic and anti-tetanus vaccine, as a precaution against infection.

Foaling complications

When to call the vet

It is important for the stud groom or mare owner to recognise when to call the vet as prompt action can save mare and/or foal.

(1) When the waters break and nothing else happens.
(2) When the waters break and one foot only shows, indicating a malpresentation or distocia.
(3) Over-large foal which gets stuck during foaling.
(4) Bleeding from vulva − could be internal haemorrhage.
(5) Broken or dislocated pelvis.
(6) Recto-vaginal fistula.
(7) Bruising and tearing of vulva (stitching and Caslick's operation).
(8) Post-natal laminitis.
(9) Mare not cleansed.
(10) Prolapse.
(11) Colic.
(12) Extended labour pains.

Foal
(1) Meconium retention.
(2) Inject 'flu, anti-tetanus and antibiotics (long acting).
(3) Deformities − contracted tendons, parrot mouth or bent limbs.
(4) Bleeding from navel.
(5) Dummy foals (brain damage).
(6) Diarrhoea, scouring.
(7) Joint ill.
(8) Entropian.
(9) Tetanus.

(10) Umbilical hernia.
(11) Pervious urachus.
(12) Ruptured bladder.
(13) Haemolytic disease.
(14) Heart problems.

Bruising and tearing of the vagina and vulva

A mare, whether a maiden or not, may easily become bruised or torn during foaling. Bruising is common, especially with a difficult foaling or the delivery of a large foal, and normally heals quickly. The area around the vulva will be swollen and inside blue—black and bruised. Any excessive bleeding at this time should be noted and the vet called, as she may be torn and need stitching. Similarly, a mare with a Caslick's operation which has been cut open prior to foaling will need to be stitched the following day. In order to give her time to heal it may be necessary to leave the foaling heat and cover her on her next season. The mare can be covered on the foaling heat even if she is stitched, if she has had an easy foaling. A course of antibiotics is usually recommended in cases of bruising and tearing.

Mastitis

This is uncommon in mares but may occur at weaning even with good stud management. Mastitis can also occur in pregnant mares and in barren mares in summer from an infection carried by flies. The mare's temperature, pulse and respiration are elevated and the udder is hot, inflamed, hard, and extremely painful. Milk from an infected mare will be thin and contain clots of pus. The pain will make the mare reluctant to let the foal drink and consequently the udder becomes even more full and uncomfortable. This infection needs immediate attention and a course of antibiotics should be given and the mare milked off until the quality of her milk returns to normal. Terramycin can be administered directly into the udder through the teat, although some feel this is futile. It is advisable to keep the mare on a non-concentrate diet and give her plenty of exercise.

Damage to nerves and pelvis (post-partum paralysis)

During the final stages of foaling when the mare is getting up and going down she may hurt herself. This is why it is important to give her plenty of good thick bedding to minimise this risk. The nerves may be damaged or the pelvis cracked or dislocated. Therapy may help the dislocation but only time will heal any damage to nerves or bone. In severe cases the mare will be unable to rise unaided and may show signs of unsteadiness when standing. If damage is suspected the mare and foal must not be left unaccompanied as the mare may fall on the foal. The administration of pain killers in some cases may help any discomfort.

Retained afterbirth

It is fairly common for mares to retain the afterbirth or placenta after foaling; this can be serious if the afterbirth is left inside the mare for more than a few hours. With veterinary help the situation can be easily rectified but, if left, laminitis or septicaemia may occur which can be fatal. Some excellent broodmares will retain their afterbirth after each foaling. The veterinary surgeon may either slowly twist the placenta which will help it detach from the uterus inside the mare, or give an intravenous injection of oxytocin which will cause smooth muscle contractions of the uterus and help expulsion of the placenta. Pessaries must be put into the uterus and a course of antibiotics given to eliminate any infection. In cases of retained afterbirth it is wise not to cover the mare on the foaling heat but leave her until her next season when her uterus should have returned to its normal size and condition.

Dystocia

This is a general term for any difficult foaling or malpresentation of the foal during birth (Fig. 6.17). After the mare has broken waters the foal should soon appear; however, if the foal has either a leg or head and neck in a bad position, nothing will happen. If left the mare will have normal contractions but will be unable to give birth unaided; she may even prolapse and in severe cases she may need a caesarean operation if she and the foal are to be saved.

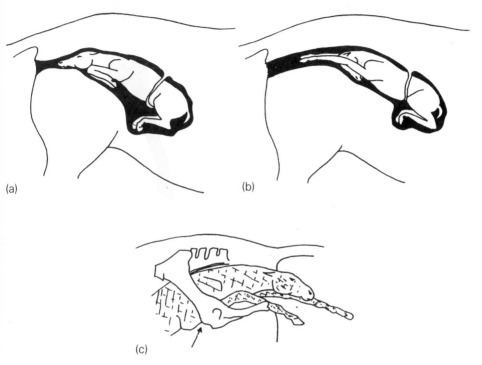

(a)

(b)

(c)

Fig. 6.17 (a) One or both forelegs may fold under the foal during delivery. In order to correct this position the foal should be pushed back into the uterus and each leg extended so that the foal is aligned correctly for delivery. While manipulating the leg the foal's foot should be covered with one hand so that the uterine wall is not damaged.
(b) When the foal's nose is not found in the birth canal it may be found tucked downwards. The foal must be pushed back so that its head can be lifted.
(c) The foal can become lodged at the pelvic brim when its forelegs are not extended properly. The flexed elbow increases the foal's diameter and hinders delivery. The foal should be pushed back until its forelegs can be stretched forward.

With minor problems the foal's position may be righted when the mare gets up and down or the stud groom can attempt to correct the problem by moving the foal's position between the mare's contractions.

In the case of a breech-birth, where the foal comes out backwards, the hind legs are seen first and the foal may be strangled by the umbilical cord. The umbilical cord may also break and the foal may suffocate. Thus it is important that the foal is delivered as quickly as possible.

Approximately 4 per cent of foals born are badly presented in some way, but these can easily be saved with some help provided that the situation is quickly recognised and appropriate action taken.

Haemorrhage of uterine arteries

This may happen a few hours after foaling and in the majority of cases there are no external signs of bleeding. The mare will show signs of colic; she will sweat up and the membranes around the eye and gums will be pale and anaemic. Eventually she will become very weak and may even become unconscious. In some cases the bleeding may stop itself but, if not, blood transfusions may need to be given. Generally the condition is fatal.

Uterine rupture

Uterine rupture rarely happens but it is very serious and can occur at any time during pregnancy, although it most commonly occurs during foaling. It may be associated with a kick or fall or the foal putting a foot through the wall of the uterus. The rupture can be treated by prompt surgery but this is extremely difficult and rarely successful.

Recto-vaginal fistula

An abnormally presented foal may push a foot through the wall of the vagina into the rectum during foaling. The vet should be called at once, but in the meantime the foal should be pushed back to correct the position so that the mare can foal normally. An operation may be carried out to mend the tear but again is rarely successful.

Prolapse

The mare's uterus may prolapse at any time up to approximately 24 hours after foaling. Prolapse is caused by excessive straining during a difficult foaling and the uterus turns itself inside out and comes out of the vagina. The uterus must be supported by means of a sheet and must be kept as clean as possible until the veterinary surgeon arrives. He will endeavour to push the uterus back inside the mare and stitch the vulva to hold it in place; an injection to stop the mare straining may also be given. A mare may also get a partial prolapse in which one horn only turns inside out. There may be no visible external signs but she will in some cases show signs of colic.

In all cases of foaling complications the signs should be immediately recognised and the vet called. It is better to have a red face and a vet's bill than a dead mare or foal!

Foal ailments

Retained meconium

Meconium is the hard black droppings which collect in the foal's bowel while it is in the uterus. These should be expelled soon after the foal is born. Failure to do so will block the foal's gut causing pain. It is common with foals which are slow to drink as the mare's first milk, or colostrum, also helps as a laxative. A foal suffering from meconium retention will stand straining with its back arched and tail up but no droppings are passed. If lying down, it may lie in a strange position with its front legs crossed over its head and it may look colicky. Normally the stud groom can removed the meconium easily by inserting a finger and easing the first droppings out. In more severe cases − and nowadays this is often a routine treatment − an enema is given. The best enemas are the ones used in hospitals for humans; they consist of plastic bags filled with a solution which will help dissolve, lubricate and pass the droppings more easily. A soft plastic tube is attached to the bag which is inserted into the foal's anus and the contents of the bag squeezed into the foal's rectum. This has spectacular results and is extremely effective. Liquid paraffin may also be given orally, or alternatively a 'home made' enema using warm soapy water, a syringe and soft plastic or rubber tubing.

As a precaution to avoid meconium retention the mare should receive plenty of exercise and a good laxative diet in the last weeks prior to foaling. A foal which is over-straining for too long may have extra complications, such as ruptured bladder and the vet should be called.

Diarrhoea

This complaint commonly occurs at approximately 10 days after foaling when the mare first comes into season. This appears to be connected with the mare's hormonal changes which affect the milk, although more recently it has been shown that the foal may be scouring from worms even at this early age. A foal may be wormed with a suitable wormer from as soon as four to five days old, which may stop the scouring; it is also advisable to worm the mare at the same time.

Any type of scouring in foals must be treated immediately by giving some kind of kaolin-based suspension orally because foals can become dehydrated very quickly. Some foals, however treated, may scour continually and will only stop after weaning. This is known as nutritional scours due to over-rich milk from mares grazing high quality grass.

Infectious scours

Infectious scours must also be treated immediately as this type can be a big killer in newborn foals and can spread rapidly. A new born foal with this complaint must be isolated as soon as possible and its old box disinfected and preferably not used at all for very young foals. Infectious scours can be caused by a virus or bacterial infection, therefore antibiotics must be administered rapidly.

One of the first signs prior to scouring will be a rise in temperature. When scouring starts a vile smell can be detected although nothing may be seen as the scours will shoot straight out backwards and disappear into the straw. Signs may, however, be seen on the walls of the stable. The foal's dock and surrounding areas should be washed and powdered or greased regularly throughout the day as the scours will burn the foal's skin causing discomfort and unsightly loss of hair. Signs of dehydration should be looked for; for example, the foal may drink water constantly. If a pinch of skin is taken and released it should return to normal immediately, if a tuck remains then the foal is dehydrated. In cases of dehydration electrolytes should be given to the foal to put back the salts lost. If prompt action is not taken the foal will become so dehydrated that it may die.

Kaolin can also be given, but a recent product called 'probiotics', which consists of micro-organisms, will replace the gut's natural organisms lost through scouring and help the foal's system return to normal.

In severe cases a blood transfusion which has been taken from its mother may be given to the foal.

Joint ill

This is an infection that enters via the navel, especially if the foal's umbilical stump has not been treated immediately after birth. Three to five days after birth the foal will become dull and listless and will be reluctant to drink from its mother. Its temperature will rise up to 105°F (41°C). The navel will either appear wet and oozy or dry, hot and swollen due to the formation of an abscess. The fetlock joints will also become swollen and painful and unless treated promptly these abscesses will burst and death will follow. Occasionally, abscesses will form on the internal organs − in this case death is rapid. The mare and foal should be isolated as soon as possible and the vet called; he will give the foal large doses of antibiotics. Box rest is essential.

The bacteria can live for some time after the foal is cured. Therefore the box should be disinfected with a strong disinfectant and not used for as long as is practically possible.

Contracted tendons

Contracted tendons occur when the tendons are too short for the cannon bone, causing the foal's foot to be permanently flexed backwards. The condition is more common in the forelegs and the degree of severity of the flexion can vary considerably. Foals can be born with contracted tendons and in severe cases a caesarian may even be needed to deliver the foal. In some cases contracted tendons occur after birth, the foal's foot can be seen to 'go over', and this can happen in a matter of hours. Severe cases may have to be destroyed but mild cases can be treated by massage, strapping and therapy, along with vitamin injections. Any strapping should be removed daily as the foal's delicate skin can easily become damaged and scarred. A small grass tip shoe may be fitted to encourage the heel of the hoof to push towards the ground, thus stretching the tendon. Exercise is essential, however pathetic the foal may appear when moving. When some foals are born the opposite condition may occur where the fetlock joints may touch the ground; this in most cases is less severe and will usually cure itself.

Dummy, sleepy, barker and wobbler foals

These descriptions all apply to foals suffering from the neonatal maladjustment syndrome. This is caused by lack of oxygen to the foal's brain during birth. The cord may break too soon, or the mare may cleanse during labour, thus stopping the blood supply which will lead to a lack of oxygen. The syndrome can affect the foal in different ways and the signs are usually noticed within two to three days of birth. The foal may appear normal initially but will then start to make a noise like a dog with respiratory disease, hence the name 'barker foal'. If lying down the foal may wave its legs around and be unable to stand properly. If the foal is able to stand, it will wander around the stable and appear blind, walking into the wall or the mare. It may also be reluctant or unable to drink due to loss of the suck reflex. The foal should be returned, as near as possible, to prenatal conditions; that is, on a blanket in a dark stable with a heat lamp to keep it warm. The vet should be called at once if permanent brain damage is to be avoided.

Entropian

This condition will become apparent soon after birth. The symptoms are chronic watering of the eyes. On closer inspection it will be found that the eyelids are ingrowing and the eyelashes are irritating and inflaming the surface of the eye. The eyelids may return to their correct position by regularly turning them back but they may need to be stitched back by the veterinary surgeon, the stitches being removed after 10–14 days.

Tetanus (lockjaw)

This takes the same form in a foal as it does in the adult horse. It can be avoided by giving the mare a anti-tetanus booster approximately one month before the foal is due. This will provide the foal with temporary immunity via antibodies in the colostrum. If this is not possible, or if the mare is very late foaling, then the foal should be injected immediately after birth, giving it temporary cover followed by a complete course at three months old. Should the foal cut itself between one and three months old then additional temporary cover should be given by means of a tetanus toxoid injection.

Snotty noses

This is a common complaint in foals up until weaning. It is best left so that the foal builds up its own resistance, but in bad cases antibiotic treatment will be needed. Having a snotty nose can pull the foal down and spread rapidly from foal to foal. Even after treatment it may quickly reappear. The foal will generally grow out of this complaint but constant observation must be kept as in some cases it can lead to pneumonia.

Umbilical hernia

This will be first noticed as a swelling in the navel region where the umbilical cord was attached. It may consist of a pocket of skin with a short length of cord inside; this may be removed by simply 'ringing' using a lamb castrating ring which, when left in place close to the stomach, will cause the cord and skin to fall off eventually. In more serious cases it may be that part of the bowel is protruding through the umbilical ring which has failed to heal in the normal way. This may need surgery, but in some cases can cure itself.

Pervious urachus

The urachus is the tube taking urine away from the foal while it is in the uterus. This will occasionally fail to close and urine leaks constantly. The urine can then cause infection and inflammation and pain. Prompt surgical treatment is therefore necessary.

Ruptured bladder

This is generally attributed to excessive pressure at birth, but may also occur following a severe case of retained meconium due to the foal's constant straining. The stomach will rapidly swell as urine leaks from the bladder and may be mistaken for meconium retention. Surgery is generally effective if the diagnosis is prompt.

Haemolytic disease

This is caused by incompatibility between the blood of the foal's sire and the foal's dam. This results in a blood disease in the newborn foal similar to human Rhesus babies. The foal inherits from its sire a blood group which is incompatible with that of its dam. A few of the foetus's red blood cells pass via the placenta into the mare's bloodstream; the mare becomes sensitised to these 'foreign' cells and produces antibodies against the red cells of her foal.

These antibodies are concentrated into the colostrum at the time of birth so that when the foal drinks the antibodies enter its blood stream and destroy its red cells. This causes severe jaundice and anaemia, which can rapidly lead to death. Haemolytic disease can be recognised by yellow membranes of the eye and gums, red-stained urine and accelerated pulse and respiration. Haemolytic disease can, however, be prevented. Blood tests are taken three to four weeks before foaling and again two weeks later. Any significant increase in the antiboay level during this time would indicate that the foal is at risk.

Any foal considered to be at risk should be muzzled for the first 36 hours and fed colostrum from another mare, the dam 'milked off' and the milk discarded. After 36 hours the foal may be allowed to drink from its mother as the foal's gut is no longer able to absorb antibodies. Should this disease not have been suspected, it will be clearly apparent within 36 hours of birth. If suspected after the foal has drunk, put the muzzle on immediately and call the vet who may treat with a complete blood transfusion.

Heart problem (failure of the ductus arteriosus to close)

In the foetus it is not necessary for the blood to pass through the lungs and a short blood vessel called the ductus arteriosus allows blood to by-pass the non-functional lungs. Within a few days of birth this blood vessel should close; failure of the ductus arteriosus to close can lead to foals being dull, weak, listless and off suck.

Very little research has been conducted on this condition; therefore there is little information on successful treatment. However, in most cases, this condition appears to cure itself within a few days.

Parrot mouth or over-shot jaw

This is an hereditary deformity in which the upper incisor teeth overlap the lower incisors. This will make life difficult for the foal as it will be unable to bite food and therefore be difficult to keep in good condition. For this reason, in severe cases the foal should be destroyed − especially if upper and lower teeth do not actually touch. If these teeth do touch this is less serious and the foal will be able to lead a normal life.

Cleft palate

The palate separates the mouth from the nasal passage and, in the case of a foal suffering from a cleft palate, is split, allowing the milk to pass from the mouth into the nasal passage. The first symptom of a cleft palate in the newborn foal is milk running from the foal's nose when it drinks. It will also make a noise similar to snoring, which will appear to come from the back of the throat. The vet will confirm this condition by using an endoscope to look at the throat and although surgery is possible it is rarely successful.

Orphan foals

An orphan foal is a foal that has lost its mother for some reason; it may be that she died at or soon after giving birth or simply that she has rejected the foal completely. She may be incapable of providing enough natural nourishment or her rejection is because, inexplicably, she believes that there is something incurably wrong with her foal. In some cases the mare is quite correct and the foal will not survive even with human intervention. The newborn foal relies on its mother for immunity and for the development of normal behaviour patterns and the loss of the mare soon after foaling or the rejection of a foal by its mother can be a serious set-back both physically and psychologically.

Immunity

Foals orphaned before they have received adequate colostrum, i.e. during the first 36 to 48 hours of life, will not have adequate immunity to disease. Some studs keep a supply of frozen colostrum, which can be given to orphan foals. If colostrum is not available a course of antibiotics can be given to protect the foal until it starts to make its own antibodies which can then fight disease. If the foal is very ill it may be given a blood transfusion which will help maintain fluid levels and also contain antibodies.

Health care of the orphan

The needs of the orphan foal are the same as those of the normal foal. However, orphans do appear to be more susceptible to disease and preventative medicine and health care are vitally important. After birth the navel should be dressed and the foal checked to see if it is normal. If the dam has died before the foal received colostrum it should be given tetanus antitoxin, antibodies and an enema. It should have its temperature, pulse and respiration taken regularly for the first week so that any abnormality can be spotted immediately. Temperature is very important and any rise above 101°F (38°C) or fall below 99°F (37°C) is an immediate danger signal. Orphan foals may scour if overfed by milk-replacers; any scouring lasting more than 24 hours is serious as foals dehydrate very easily. The foal's box should be airy and light with an infra-red heat lamp directed so that the foal can move in and out of the heat and remain comfortable. A foal rug or an old sweater may be used if the foal is still cold.

Feeding the orphan

It may be possible for a stud or owner to rear an orphan foal but if the time and facilities are not available it can be sent to the National Foaling Bank where it will receive experienced care and attention. Orphans can be bottle-fed but the ideal is to foster them on to another mare.

(a) Foster mothers
A foster mother may be a mare who has lost her own foal, a milky mare who will accept another foal or a mare whose foal is old enough to be weaned.

Great care must be taken when introducing the orphan to the foster mother; she is more likely to accept it if it smells familiar. This can be done by placing her dead foal's skin over the orphan, rubbing the foal with the foster mother's milk or manure or putting a strong-smelling substance on the mare's nose to mask the foal's smell. The mare can be put in stocks, have a leg held up or be twitched while the foal is encouraged to suckle. The mare should be allowed to smell the foal but must be watched carefully in case she becomes aggressive. This should be repeated every two to three hours. As the mare begins to accept the foal she can be allowed to stand on her own and gradually allowed more freedom until they can be left together. It is unlikely that the mare will reject the foal once the initial adoption takes place but both mare and foal should be closely watched for the first few days after they have been turned out together.

If the mare has a foal at foot the orphan foal can be kept separated by placing straw bales across a large box, so that both the mare and foal can see the orphan and become accustomed to it. The orphan can be let in to the mare every two hours to suckle. A careful eye must be kept on the natural foal as it may become aggressive and it may be necessary to sedate it initially. This method requires a 24 hour watch on the mare and foals for possibly as long as three days. However, it is well worthwhile as the orphan foal receives discipline and nourishment from its new mother and can lead a fairly normal life. It may be necessary to supplement the orphan foal's diet with mare's milk replacer. The foal should be given milk in a bucket and left to drink rather than suckle from a bottle which can teach it bad habits. As the foal matures creep feed can be introduced into the milk to encourage growth and make weaning easier.

(b) Bottle feeding

If a foster mother is not available the foal must be hand-reared; a feeding bottle can be made by fitting a clean soft-drink bottle with a lamb-teat. The foal should be taught to drink milk from a bucket as soon as possible; allowing the foal to lick milk off the finger and then immersing the finger in the bucket will encourage the foal to follow the finger and discover the milk. It should never be forced to put its head in the bucket.

If the foal is premature or lacks a normal suck reflex it must be fed through a stomach-tube until it has learned to suckle. A soft rubber tube is thoroughly cleansed in hot water, greased and then passed through the foal's nostril, down its throat and into the stomach. Milk is then slowly poured into a funnel attached to the tube. The foal's nostril should be greased and when the tube reaches the back of the throat the foal should swallow so that the tube passes down the gullet, *not* the wind-pipe. Before putting any milk down the tube the handler should listen to the end of the tube carefully; if the tube is in the lungs a characteristic noise will be heard, and the tube should be slowly withdrawn and then replaced into the stomach. A vet must show the foal handlers how to use the stomach tube correctly.

Foals that are hand-reared can become too dependent on human beings and try to play as they would with other foals. They must be firmly disciplined − a fun game with a foal can become dangerous with a yearling!

A reputable mare's milk replacer made up as directed should be used; cow's milk is not suitable although goat's milk can be used. During the first two weeks normal, healthy foals should be fed every two hours and during the second two weeks every four hours. They should then be fed four times a day until weaning. Individual needs as to amount and frequency of feeds will vary according to age, size and state of health but generally foals should be fed about quarter to half a pint at each feed initially. Each week the amount should be increased to the maximum that the foal will happily eat without scouring; scouring without fever or other signs or illness may indicate that the foal is being overfed and its diet should be adjusted accordingly.

(c) Creep feeding

The orphan foal should be introduced to solid feed as quickly as possible. This will encourage gut development and allow the foal to be weaned from the bucket earlier. Pellets containing milk powder or milk replacer pellets can be added to the bucket or liquid milk replacer which will encourage the foal to eat them. The pellets should contain at least 16−18 per cent high quality protein, have a high energy content and adequate levels of minerals and vitamins, particularly calcium and phosphorus.

Weaning

The hand-reared orphan foal should be 'weaned' from milk replacers as soon as possible; early weaning will save labour costs and allow the foal to lead a more normal life. The foal must look well and be eating adequate creep feed, the milk ration should be gradually reduced and creep feeding continued for another three months.

7 Weaning

Weaning is the separation of the foal from its dam and its change in diet from milk to grass.

Preparing the foal for weaning and the ways of easing stress in the foal

Eventually, the day will arrive when the foal has to be parted from its mother and with good management some of the stress and worry can be avoided. It is very important to prepare foals before weaning and there are several factors to be taken into consideration.

It is essential that the foal should become independent enough to be eating concentrates on its own. The best method to use, especially when the mare and foal are stabled, is to creep feed. This can be done in several different ways; possibly the most successful and practical method is to use a specially constructed manger. The manger is similar in size to an automatic water drinker and is divided centrally by means of a movable bar which may be adjusted so as to enable only the foal to eat from it. The mare will be unable to push her muzzle through the space left by the bar and therefore controlled quantities of concentrates can be given to the foal. Alternatively, a bar or beam can be fitted into the stable at a suitable height to allow only the foal to pass under it enabling access to food placed in the other half of the stable. Similarly, a corral or pen may be constructed in the paddock so that the foal can be fed in the same way. Although in most cases the foal will go off its food for a short period after weaning it will regain its appetite more quickly if it has been used to eating concentrates prior to the separation.

Colt foals, if not to be kept entire, may be castrated prior to weaning at approximately four to five months old. Weight loss is considerably lessened if they have been creep fed prior to castration.

It is often believed that leaving a colt entire for two to three years until castrating will help in its development but the inconvenience of having a boisterous young colt to manage completely outweighs any extra physical development that may occur should the animal be cut later.

It is vital that the foal should have had its influenza and anti-tetanus inoculations well before the weaning date. As in the adult horse, a foal should not sweat within the next few days as this may affect its health. It is important that the foal has been regularly wormed since birth. A foal in poor condition at weaning will suffer unnecessarily as will a sick one; any foal with a viral infection, a cough or snotty nose should not be weaned until it has recovered.

It is practical to wean foals in groups or pairs as the foals can be stabled in close proximity to each other and the mares can be put into paddocks together, thus eliminating the danger of animals being injured through trying to be reunited. For the one mare and foal owner the foal could be accompanied by an older, quiet mare or gelding. In some cases a local mare owner might be in a similar situation and one or the other could accommodate either both mares or both foals. Donkey companionship should be avoided unless the donkey is completely free from lungworm.

When to wean

As a guideline, most stud owners when weaning in groups will wean their foals at approximately six months. This will vary according to several factors. For example, the mare may not be producing enough milk to sustain the foal. In this case it would be better to wean much earlier and put the mare onto poor quality grazing if she is fat, especially if she is not in foal. The foal could then be fed extra concentrates including some form of milk powder or pellets. If the mare is barren the foal could be left on her until the following spring and weaned when she returns to stud. In this case only one stable is occupied during the winter.

If the mare is in foal and losing condition because of feeding a large robust foal they could also be separated earlier. Foals required for the autumn sales or for the shows the following year should be stabled at nights and fed regularly. In some cases these foals should be rugged as a high standard of production of the young animal is vital for success at either of these.

Types of weaning

(i) Sudden or abrupt weaning

In this method mares and foals are separated and the mares are put onto poor grazing out of earshot of the foals. The dramatic drop in nutrition from good to bad grazing encourages the mare to take more exercise when searching for food and she should 'dry up' within a few days. After this short period the mares will be able to cope with a higher plane of nutrition, thus the risk of weight loss is minimised. For safety reasons it is essential that the foals are stabled, and it is necessary to prepare the foal's stable prior to separation, ensuring that there is nothing that the foal can injure itself upon. Foals often become violent and crash around the stable thus injuring themselves; the water bucket should be removed or put into a holder and hay nets should be removed and hay fed in racks or small quantities on the ground to avoid wastage. A small feed should be placed in the manger which will help to occupy the foal and avoid stress. Weaning boxes must have either a finely meshed grid or a strongly built top door; a foal when left alone becomes oblivious to danger and serious injury can be sustained should it try to jump out.

At the time of actual separation mares should be led in groups directly to the paddock; bridles should be used as headcollars may be unsafe if the mare becomes difficult, and the foals should be stabled separately with the top doors closed. Weaning in pairs can only cause complications – eventually they will have to be separated and this can be as stressful as being separated from the mare. Also one foal may be bullied when feeding or one may eat more than the other.

(ii) Gradual weaning

This is when the mare and foal are separated for part of the day or night within sight of each other. The mare will dry up gradually with less pain, but eventually the time will come when they have to be separated totally which may be equally as stressful as the sudden weaning method.

(iii) Paddock weaning or pied piper

This can only apply to an established group of mares and foals. The mares will be removed one by one until eventually only one barren mare will be left with the foals to act as nanny. Although a practical method of weaning it does mean that controlled feeding of the foals is difficult.

Care of mare and foal after weaning

Foal

When the foals have settled down the top doors can be left open, providing a grid is present. Do not turn out the foals into a paddock too soon as although the foal may appear settled it could easily become lonely and rush around looking for its mother. It is unwise to turn a foal out on its own straight after weaning as it could injure itself. Colt foals, if they are to be kept as stallions, will have to live on their own eventually and they must start to become accustomed to this from weaning. Mature stallions if turned out together will eventually fight.

The daily routine should consist of teaching the foal to tie up, picking out its feet, grooming and being led out in groups and on its own.

Mare

The mare must be checked daily for signs of mastitis which may occur in high milk yielding mares, however strict the management. Mastitis is the inflammation of the udder caused by bacterial infection and the udder becomes extremely swollen, causing great discomfort. This should be treated by regular 'stripping' or milking of the udder. The milk will look as if it is curdled − in fact, pus is present. A course of antibiotic injections should be administered daily and in some cases terramycin ointment inserted into the teat. Failure to correct mastitis could lead to the loss of use of all or part of the udder.

Keeping a colt entire

Most breeders are happy when a mare produces a colt foal rather than a filly, because it is claimed that geldings are often easier to manage and make better riding horses or hunters. In these cases a colt foal is cut at approximately five to six months whilst still on the mother and weaned later. Nowadays it is becoming more fashionable to compete cross-bred and pure-bred colts and stallions. The management of entires, especially if they are covering mares, is certainly more difficult than managing geldings and it must be carefully considered whether or not to keep a colt foal entire.

What to look for in a colt foal

A colt foal that is going to be kept to stand at stud must have some outstanding qualities; he should be bred well enough to attract potential mare owners. His parents should either have raced or competed successfully themselves or have already produced proven stock. The foal must have exceptional presence; his ears pricked, neck arched and he should stand square naturally. If a human walks into his stable or field then the colt should show interest at once and behave in a confident manner. He should move well enough to satisfy the requirements of the job he is doing or the stock he is likely to produce, but it must also be remembered that successful competition horses are not always the best movers. Even at a very young age a foal will show signs of its ability; outstanding movement, natural signs of jumping in the paddock, even jumping out of the field must indicate that the foal can jump and is bold. Some breeders loose jump foals or yearlings over very small obstacles in an indoor school. Temperament, ability and movement may be easily assessed at this age. A stallion should have good conformation and pass this on to his stock but, remember, the perfect horse does not exist!

Handling the colt foal

Depending on his temperament a colt can run with other foals for some time, but if he is to be kept entire then at some stage he should learn to live on his own, as this is how he will spend the rest of his life. First he must be stabled separately and taught to tie up in the box. He must be made to stand still and have his feet picked out and trimmed. Normally a colt will be used as a stallion at three to four years old although in some cases his stud duties may start sooner or even later. He should be led out and disciplined in hand and perhaps turned out in an indoor school or stallion paddock. He must live as normal a life as possible to prevent him acquiring bad habits such as windsucking and crib-biting or any other vices.

There are many ways of restraining a boisterous colt when leading him in or out of the stable; it is possible that a good-natured colt can be led in a simple headcollar. Putting bits in a young horse's mouth is not advisable if he is to be used for competition work, as it could encourage mouth problems at an early age. It is better to use a chain over the horse's nose from a headcollar and a stick should always be carried. If a young colt has been well handled and disciplined from an early age, when he becomes a mature stallion he will be easier to manage.

He should start the stud season in good physical condition, and as with the mare, may receive extended periods of artificial light. He should either be ridden daily, led out in hand or lunged. Loose working if suitable facilities are available is an excellent way of keeping the stallion happy and contented. The stallion must have been swabbed for any contagious diseases which may affect his fertility. This should be done in January or February so that any infection can be treated before the start of the covering season.

8 Nutrition

Art or science?

Many people are undeniably better at feeding horses than others; a good stockman can get horses to 'do better' by an inherent ability to appreciate the needs of his horses. This art comes into the 'doing', i.e. actually feeding horses on a day-to-day basis. The science can be thought of as 'knowing', i.e. an understanding of nutrient requirements, deficiency symptoms and feed constituents.

Unfortunately many of our feeding practices are based on knowledge gained from other animals and it is questionable whether these are applicable to horses. This situation has arisen because, compared to sheep, cattle and pigs, horses are too expensive to provide in the numbers necessary for a valid feed trial. The Government does not yet recognise the horse as being important enough in agriculture to provide much funding for research.

The situation is also confused because, unlike agricultural animals, the horse has been selected for performance, not liveweight gain or milk production. Nutrient requirements for performance are much more difficult to gauge.

It is helpful, however, for the stud groom, mare owner and youngstock owner to have a basic knowledge of:

(a) the anatomy of the horse's gut;
(b) nutrient requirements;
(c) feed constituents.

This knowledge can be used to feed horses correctly and economically.

The anatomy of the horse's gut

The function of the gut is to digest food, i.e. break down food to make nutrients available to the body for the production of energy, building of muscle and bone and the storage of fat. This break-down is achieved by enzymes and micro-organisms.

Enzymes are chemicals produced by the stomach and small intestines (Figs. 8.1, 8.2 and 8.3) which eventually reduce soluble carbohydrates to simple sugars such as glucose, proteins to amino acids, and fats to fatty acids and glycerol. These end-products are absorbed across the gut wall and utilised by the body.

Micro-organisms such as bacteria and protozoa are concentrated in the horse's hind gut (caecum and colon). Their task is to break down the fibrous part of the diet. Fibre is composed of insoluble carbohydrate and is metabolised by the micro-organisms to release volatile fatty acids which can be used by the horse to provide energy. These micro-organisms rely upon a constant environment and any sudden change in diet which affects this environment can kill them.

The horse differs from the dog or pig in that most digestion takes place in the hind gut because the majority of the diet is fibrous, therefore taking longer to digest. The horse is also unlike grass eaters such as the cow or sheep, in that it has only one

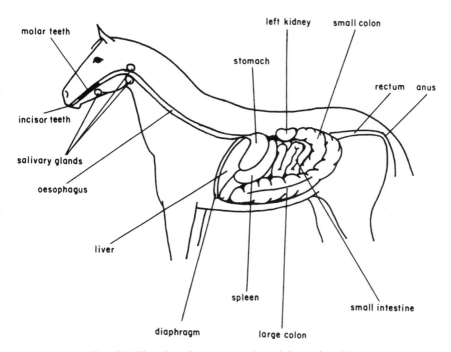

Fig. 8.1 The digestive system viewed from the side.

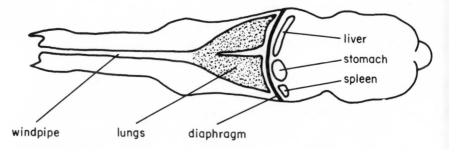

Fig. 8.2 The relationship between the stomach, diaphragm and lungs.

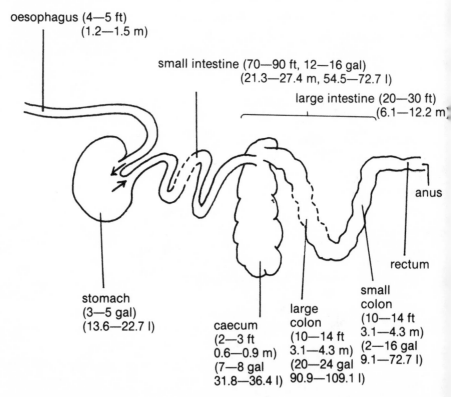

Fig. 8.3 The alimentary canal of the horse, showing dimensions and capacities of different regions.

stomach whereas the ruminant has four stomachs; it is here that microbial digestion of fibre takes place. This leaves the horse having a gut that resembles that of a rabbit more closely than anything else!

Several of the tried and trusted 'rules of good feeding' can be explained by looking at the anatomy and physiology of the horse's gut:

(a) *Feed little and often*: small feeds do not overfill the relatively small stomach and feeding frequently keeps the large hind gut topped up and gives the micro-organisms a constant environment.

(b) *Make any changes gradually*: sudden changes in diet, e.g. grass to hay and concentrates, can cause digestive upset. Different micro-organisms break down different feedstuffs and changes in diet may deprive certain bacteria of food and they die releasing toxins which can cause diarrhoea.

(c) *Water before feeding*: if the horse drinks large amounts of water immediately after a feed, partially digested contents can be washed from the stomach into the small intestine and may cause colic. Generally speaking, horses should have fresh clean water available at all times.

Feeding the broodmare

The pregnant mare

The pregnant mare should be kept healthy and in good condition, not fat. During winter or summer the mare's ribs should not be seen but should be detected on touch with no appreciable layer of fat. This will make foaling easier and enhance her chances of conceiving again. If she is in good condition the mare should be fed for maintenance during the first eight months of pregnancy as the growing foetus is not making any appreciable demands upon her. In other words, assuming she conceives in April, grass keep will be adequate until September or October depending upon the availability of pasture and providing she is still in good condition. As the grass supply wanes she should be fed good quality hay until Christmas; after this she should be stabled at night and concentrate feeds introduced. During the last three months of pregnancy her nutrient requirements increase. At the same time the foetus occupies an increasing proportion of her abdomen, thus her capacity for bulk feed may drop and must be made up by feeding more concentrate feed with a protein content of about 16 to 17 per cent. Some mares may naturally go off their feed as foaling approaches; this is nature's way of saying that she is unable to cope with this high energy and protein diet, and an increase in good quality hay or lucerne plus appetising succulents in the feed should provide all necessary nutrients. In later foaling mares, the flush of spring grass will have the same effect. Feeding in this way may help reduce the problem of meconium retention in the newborn foal.

It is important not to overfeed the broodmare; it is a fallacy that only ponies are susceptible to laminitis — overfed barren and pregnant mares are equally at risk.

There is conflicting evidence about the effect of overfeeding or underfeeding during the last three months of pregnancy. Generally speaking, the mare foaling in fat condition should maintain, not lose, weight and thin mares should gain weight. Any imposed weight reduction in obese pregnant mares should take place before the last three months of pregnancy and preferably before they are covered.

It is important that the diet is balanced in calcium and phosphorus as imbalance may lead to a weak foal (Table 8.1).

In the 24 hours before foaling, the mare should be fed good quality hay and low energy concentrates, e.g bran and horse and pony nuts. It is likely that the mare will go off her feed anyway at

Table 8.1 The minimum daily requirement of calcium and phosphorus.

Age	Body weight (kg)	Calcium	Phosphorus	Limestone (g/day)	Dicalcium phosphate (g/day)
3	100	37	31	104	148
6	200	33	27	92	108
12	300	31	25	87	92
18	375	28	23	78	72
mature	450	23	18	64	
lactating	500	33	235	92	94

Note: One ounce = 28 grams.

this time. The first feed after parturition may be the controversial bran mash as this is appetising and easy to eat; subsequently the mare will go on to a higher plane of nutrition.

The lactating mare

The nutrient requirements of a lactating mare increase dramatically, being equivalent to those of a fit horse hunting regularly. These nutrients are transformed into milk with the mare reaching peak lactation at about 30 days post-partum and yielding about 12 kg per day during early lactation and 10 kg per day during the second half of her lactation. The first milk, colostrum, has a very high protein content as it contains antibodies which give immunity to the newborn foal. As previously mentioned, it is essential that the foal suckles and receives this immunity during the first 24 to 36 hours of life; the gut wall will not allow antibodies to pass into the bloodstream after this time.

The lactating mare needs good quality hay and high protein and energy concentrates, particularly if she is foaling before the spring grass is available to supplement her diet. The calcium content of the diet should be watched and the mare should receive about 92 g of limestone per day (Table 8.1). Low calcium intake may result in milk fever; this is a condition mainly associated with high yielding dairy cows and is caused by a sudden draining of blood calcium. Milk fever occasionally occurs in mares but can be avoided by feeding about 40 g of limestone during the last two or three weeks of pregnancy and about 30,000 IU vitamin D per day (equivalent to 10 ml of cod liver oil).

As the lactation progresses and the mare has increasing access to grass she is unlikely to require any supplementary feeding unless she is in poor condition, being prepared for the show ring or if grazing is limited.

Barren and maiden mares

Providing the mare is neither too fat nor too thin, a similar feeding regime to that of the mare in early pregnancy can be followed, i.e. a maintenance ration of good quality forage until Christmas. At about Christmas time, the mare should be stabled at night, rugged and fed a concentrate feed once or twice a day. The idea is to persuade the mare's body systems that Spring is on the way and thus start her oestrous cycles earlier in the year than normal. An increase in temperature (rugs), a rising plane of nutrition (concentrate feed) and, most importantly, increased day length imitate Spring-like conditions. The mare should have a programme of 8 hours dark and 16 hours light provided by leaving a light on in her stable for the required length of time − easily done in practice by using a time switch. Around Christmas-time, for example, the light should be on from 4 pm until about midnight. The light acts through the eyes on the part of the mare's brain that initiates hormonal activity and stimulates the pituitary gland to produce increasing levels of FSH and LH, so that the mare should be regularly cycling by the end of February or beginning of March.

Feeding youngstock

The newborn foal

It is vital that the newborn foal receives adequate colostrum and suckles normally. Foals suckle very frequently, up to 100 times in 24 hours during the first week; small amounts of milk taken frequently are less likely to cause digestive upsets. By 10 to 21 days the foal should be nibbling hay, grass and concentrates and if the foal is being prepared for the show ring or for early performance like flat racing or if the mare is short of milk, then the foal should be creep fed. Creep feeding is giving the foal special concentrate feed not accessible to its dam, and it accelerates growth and the development of the digestive tract so that weaning is less traumatic. A creep feed based on skimmed milk (18 per cent crude protein) can be used from two weeks onwards. From ten to fourteen weeks the feed should be gradually changed to a growing foal diet or stud cube (14 per cent crude protein). Milk pellets should not be fed prior to weaning as the gut must be taught to digest other nutrients.

Foals that are growing well do not need a creep feed until about eight weeks before weaning. The object of this is:

(a) to compensate for the mare's falling milk yield;
(b) to compensate for the decline in pasture quality;
(c) to accustom the foal's gut to concentrates.

The creep feed eaten will probably be about 0.5—0.75 kg per 100 kg body weight. To put this in perspective, a foal likely to weigh 550 kg at maturity (about 16 hh middleweight) will weigh about 150—200 kg at weaning and be eating about 1 to 1½ kg of concentrate feed.

Foals can suffer from growth-associated problems such as epiphysitis and contracted tendons if they are overfed. If these problems arise then both mare and foal's concentrate feed should be cut out for three to four weeks. This restriction will not affect the foal's mature weight if it is carefully regulated. Foals that are born particularly high on their toes should be turned out to grass with their dams as soon as possible and not allowed to grow too quickly.

Epiphysitis is a condition that is usually found in the fetlock joint at the end of the cannon bone and in the knee at the bottom of the radius. If the condition is severe the foal should be box-rested and only fed good quality hay until the 'bumps' go down. This allows the joints to mature without excess pressure due to the foal being too heavy. Where the condition is slight it will probably right itself.

Contracted tendons is a condition associated with foals that are doing well on milky mares that are grazing good grass. The tendons do not actually contract but do not appear to keep up with the rate of bone growth: the heel will rise and the front of the hoof becomes slightly concave. As soon as the condition is seen, the foal must have a reduced diet, exercise and weekly rasping of the heels.

Weanlings

The growth-related problem of epiphysitis can occur in weanlings and yearlings and so it is important that they receive a balanced and adequate diet but are not overfed during their first winter.

It is useful to feed a reputable stud cube designed for young growing horses along with good quality hay during the winter. It is unlikely that a weanling would require more than 3½ kg of concentrates if allowed as much hay as it wanted. Unless the weanling is being prepared for sale or show it should be possible to feel a rib with no appreciable layer of fat.

The growing horse should always have its diet supplemented with calcium and phosphorus to ensure good bone growth.

Weanlings may be yarded during their first winter. This allows them plenty of room for exercise while being warm and sheltered but a watch must be kept for bullying at feed time.

Feeding the stallion

Out of the breeding season, a resting stallion in good condition should maintain his condition on good quality hay and horse and pony nuts. This plane of nutrition should be gradually increased after Christmas as exercise is introduced to fitten him before the covering season. During the covering season stallions may become difficult to feed, going off their feed and losing condition. They should be fed a stud cube or equivalent concentrate mix and good quality hay at the rate of 0.75 to 1.5 kg per 100 kg body weight, i.e. 3.75 to 7.5 kg for a 16.2 hh thoroughbred of 500 kg. Generally speaking, the stallion should be fed as if he is doing moderately hard work with good quality palatable feed and a suitable general mineral and vitamin E supplement. Whilst a deficiency of vitamin E has been implicated in fertility problems, feeding extra does not guarantee extra fertility!

General guidelines for good feeding

(1) Feed good quality forage.
(2) If forage is not of good quality make good the short fall by feeding extra concentrates.
(3) Do not rely solely on pasture: its quality can vary dramatically depending upon time of year, sward quality and soil nutrition. Compensate for pasture inadequacy with concentrates.
(4) Constantly check condition and growth rates to avoid poor conformation and epiphysitis.
(5) Feed adequate calcium and phosphorus in the correct ratio and quantity (Table 8.1).
(6) Ensure the use of one good general mineral and vitamin supplement; the use of more than one supplement may imbalance the ration further as they may not complement each other.

Ration formulation

It is very difficult to advise how to feed any individual horse without observing its condition, pasture and system of management, consequently Table 8.2 is only a guide based on the mythical 'average horse'.

Table 8.2 Ratio of concentrates to hay.

Type of horse	DE (MJ/kg)	Good hay (10 MJ/kg)	Medium hay (8 MJ/kg)	Crude protein %	Ca %	P %	Vitamin A IU/kg	Ration Hay (kg) (8 MJ/kg)	Ration Compound kg 11 MJ/kg 16% CP
								(lbs in brackets)	
Breeding stallion	12	45:55	55:45	12	0.45	0.35 (1.3:1)	2450	5 (11)	5 (11)
Pregnant mare (last 3 months)	10.5 =light work	25:75	35:65	11	0.5	0.35 (1.5:1)	3400	6 (13)	3 (7)
Lactating mare (first 3 months)	12 =medium work	45:55	55:45	14	0.5	0.35 (1.5:1)	2800	6 (13)	$6\frac{1}{2}$ (14)
Lactating mare (second 3 months)	11	30:70	40:60	12	0.45	0.3	2450	8 (18)	$3\frac{1}{2}$ (8)
Weanling (6 months)	13	65:35	70:30	16	0.85	0.6 (1.5:1)	2000	2 ($4\frac{1}{2}$)	4 (9)
Yearling (12 months)	12	45:55	55:45	13.5	0.55	0.4 (1.4:1)	2000	$3\frac{1}{2}$ (8)	$3\frac{1}{2}$ (8)
Two year old	11	30:70	40:60	10	0.45	0.35 (1.3:1)	2000	5 (11)	$2\frac{1}{2}$ ($5\frac{1}{2}$)

Note: Salt, bonemeal or limestone and cod liver oil should also be fed.

Digestible energy (DE) measured in megajoules (1 megajoule = 4187 Calories) per kilogram is the energy contained within a feed or diet that is available to the horse. Rations can be calculated just as they are for human beings in Calories − counting to provide the correct amount of energy for maintenance, work, lactation or growth. Good hay can contain as much as 10 MJ DE per kg, medium hay 8 MJ DE per kg while oats contain up to 14 MJ DE per kg. Obviously, if good quality hay is fed, fewer high energy concentrates are needed than if medium quality hay is used. The greater the energy requirement of the horse the more concentrates need to be fed, e.g. lactating mares and growing youngsters.

Protein is required to build tissue and so requirements are greater in horses that are producing milk or growing. Thus the lactating mare needs 14 per cent crude protein (CP) in the diet compared to 9 or 10 per cent CP in the adult resting horse. As even good quality hay does not contain this much protein it is important that the protein is supplied in the concentrate feed; most compound feeds designed for studs have 14−16 per cent crude protein. Crude protein indicates the nitrogen contained in the feed; not all the nitrogen is present as protein so this is a rather inaccurate indication of the digestible protein content. Proteins are made up of building blocks called amino acids; the amino acid lysine is frequently deficient in horse diets and when choosing a supplement for young-stock lysine should be looked for in the formulation.

Calcium and phosphorus are vital for correct bone growth and development. As well as being present in the diet in the correct quantity they must also be in the correct ratio; too much phosphorus can upset this ratio and cause symptoms of calcium deficiency. Few supplements contain adequate amounts of calcium and phosphorus and the diet should be supplemented with bonemeal or limestone.

It is essential that adequate amounts of minerals and vitamins are supplied in the diet; cod liver oil will provide the fat-soluble vitamins A, D and E and a reputable general purpose mineral and vitamin supplement will provide the rest unless a specific problem has been identified.

9 Infertility in the mare and stallion

The reproductive efficiency of both mare and stallion is vital to the success of a breeding programme and the survival of a public stud.

The mare

The complexity of the mare's reproductive role subjects her to greater physical stress and hence more potential fertility problems. These problems can be minimised through proper management, and broodmare management is a key factor in determining the success or failure of a breeding programme. Fertility problems can fall into three major categories:

(i) management problems;
(ii) physiology;
(iii) disease.

Management problems

There are two major 'man-made' factors which may reduce fertility:

(1) *The natural versus artificial breeding season*
Studies of ponies running wild have shown fertility rates of 95 per cent yet average figures for managed horses are between 50 and 70 per cent. As she evolved, the mare developed a natural breeding season during the summer months so that she subsequently foaled when the weather was warm and the grass nutritious and her foal had an optimum chance of survival. However, the thoroughbred has an arbitrarily defined breeding season extending from 15 January to 15 July with all foals automatically becoming yearlings on 1 January. Thus the imposed breeding season ends before the mare's natural breeding potential reaches a peak (Fig. 9.1). Man-

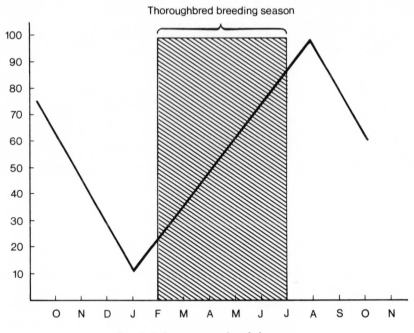

Fig. 9.1 Oestrous cycle of the mare.

agement of the mare and manipulation of the oestrous cycle have become crucial in ensuring good conception rates.

(2) *Selection*

Horses are generally selected on the basis of conformation and athletic ability not on breeding capacity. Subfertile mares have been bred from using artificial methods or veterinary intervention, resulting in inherited reproductive problems being passed on and the reproductive capacity deteriorating.

Other factors

- Nutrition: The mare is entirely dependent on us for her feed and water – correct nutrition is essential for conception, the maintenance of pregnancy and the correct development of the foetus. This will be discussed in a subsequent chapter.

- Detection of oestrus and time of breeding: Thorough teasing to detect the in-season mare and covering just prior to ovulation are management factors which will also help conception rates.

Physiology

The oestrous cycle of the mare is a complex process involving a delicate balance and interaction between the reproductive hormones. Frequently the cycle will be irregular and not conform to the ideal.

'Normal' variations
These result from normal seasonal changes during the year.

- Longer dioestrus: An abnormally long interval between successive seasons occurs if the pituitary gland is producing little or no FSH and thus there is little follicle development. If this occurs for several months during the winter it is called the winter anoestrus and accepted as normal.
- Longer 'coming in' period: Prior to being in season the mare has a 'coming in' period where she shows some but not all of the signs of accepting the stallion. This period usually lasts about 12 hours but following winter anoestrus the first attempts to come back into season are usually prolonged as the growth of follicles is slow and the mare may show 'coming in' behaviour for days.
- Longer oestrus: Oestrus may be prolonged in the spring if the pituitary gland is producing insufficient amounts of LH to cause maturation of the follicle and ovulation.

These events reflect the changeover from having relatively inactive ovaries in winter to having regular oestrous cycles in summer and generally occur in February, March and April. The changeover can be hastened by supplying artificially increased daylight hours or hormone treatment, most commonly by feeding a progesterone analogue for 10 to 15 days. In very simple terms, this allows the insufficient release of FSH and LH from the pituitary to build up while the mare is being given the progesterone so that when the treatment is withdrawn the 'flood' of hormones released is sufficient to stimulate the ovaries and for the mare to come into season. The treatment appears to be more effective if immediately followed by an injection of prostaglandin although oestrus behaviour may not be seen for between 8 to 18 days after the last dose.

Abnormal variations

- Prolonged dioestrus: Normally the corpus luteum is spontaneously formed after ovulation and ceases to function and begins to regress 14 days later due to the release of $PGF_{2\alpha}$ from the uterus. How the uterus knows when to release $PGF_{2\alpha}$ is not known but occasionally the system breaks down, $PGF_{2\alpha}$ is not released and the corpus luteum persists. This means that the mare does not come back into heat at the expected date due to the high levels of progesterone in the blood. Prolonged dioestrus may occur in dry and lactating mares and in any month of the year.

Injection of prostaglandin causes regression of the persistent corpus luteum and is safe and easy to use. However, certain brands may cause some mares to sweat heavily; dilution of the injection can avoid this but effectiveness may be reduced. The injection can be used to reduce the length of a normal period of dioestrus but can only be used if a functional corpus luteum is present, i.e. the injection is only successful if used four to five days after ovulation; the mare should come back into season two to four days after injection but can take several days longer.

Uterine irrigation with saline will also cause regression of the corpus luteum but is believed to be less effective than PG and more likely to cause uterine infection.

The most common reason for a mare not coming back into season during the breeding season is pregnancy. This must always be considered before PG treatment as this may cause abortion.

Foal heat
The mare will come into season approximately five to eight days after foaling. Covering mares during the foaling heat is controversial; on the one hand it may get the mare in foal quickly but on the other conception rates are low. The single most important factor that determines whether or not a mare will conceive during her foal heat is the condition of her reproductive tract after foaling. Even a normal foaling causes a great deal of trauma which may not have resolved itself before the foal heat, making the uterus incapable of supporting conception.

Silent heat

Nervous or shy mares or mares with foals may not show to the stallion even though they have a ripe follicle within the ovary. This is a behavioural problem that needs handling with tact and careful observation: for instance, some mares may only show to other mares in the privacy of their own box.

Lactational dioestrus

This is a period of complete absence of sexual behaviour that sometimes affects mares before or after their foal heat. This is thought to be due to a persistent corpus luteum.

Granulosa cell tumours

Ovarian tumours or cysts frequently occur in the mare. Granulosa cell tumours are the most common. They are derived from the cells that line the follicle causing the ovary to become enlarged and honeycombed with cyst-like spaces containing blood-stained fluid. The tumour produces oestrogen which inhibits the pituitary and prevents normal sexual behaviour resulting in dioestrus, nymphomania or aggressive stallion-like behaviour. Successful treatment usually involves surgical removal of the affected ovary.

Multiple ovulations

The mare's uterus is not designed to hold more than one developing foetus. Twin pregnancies result in crowding and the abortion of one or both embryos or the birth of two weak undersized foals. Twin ovulations are therefore considered undesirable and it is best to avoid breeding from a mare if a high chance of twin conception exists.

The use of the scanning machine for early detection of pregnancy has helped reduce the number of twins by allowing the vet to pinch out one conceptus or abort the mare at an early stage. However, the tendency towards twinning is believed to be hereditary and breeding from these mares may exacerbate the problem.

Reproductive diseases

Uterine infection

The uterus undergoes a continual cycle of covering, pregnancy and exposure to infection and disease which results in repeated damage and a gradual deterioration of the uterine walls. Disease of the uterus and reproductive tract not only impairs the mare's ability to conceive but may also reduce her resistance to subsequent infection. Disease-causing bacteria may be normally present in the mare's reproductive tract and development of uterine infection depends on the mare's natural resistance and the types and numbers of organisms present. Young, healthy mares tend to be less susceptible to uterine infection than older mares which have had several pregnancies. As mentioned previously, poor vulval conformation leading to contamination and wind-sucking will make mares more likely to develop infection.

Infection of the uterus leads to metritis or inflammation of the uterus; endometritis is inflammation of the endometrium which lines the inside of the uterus. The uterus becomes swollen and may contain pus and develop scar tissue. This leads to failure to conceive and a shortened oestrous cycle due to premature regression of the corpus luteum. If a mare's cycle changes from 21 days to 16 days or less endometritis should be suspected and veterinary help sought.

Acute metritis causes severe damage and is characterised by irregular oestrous cycles and the release of a creamy white fluid from the vagina during oestrus. This fluid may mat the mare's tail and run down her hind legs.

Chronic uterine infection (pyometra) is a failure of the tract to fight infection resulting in long-term and highly damaging disease. These chronic infections are often disguised as the mare appears to be very healthy and there is rarely a discharge from the vagina.

Although infections have been observed in all parts of the mare's reproductive tract the ovaries, the Fallopian tubes and cervix appear relatively resistant to infection.

Infectious organisms

Identification of the organism causing the uterine infection can be important in deciding how to treat the infection. Samples of bacteria-filled fluid (swabs) are taken from the vagina, cervix, uterus and possibly the clitoral fossa and the bacteria are cultured in the lab. A diagnosis can then be made and an appropriate treatment chosen.

- *Streptococcus*: These bacteria are commonly found on the skin, mucous membranes and intestine. *Streptococcus zooepidemicus* can cause severe inflammation of the cervix and uterus possibly resulting in abortion. Generally, streptococcus organisms play a very important role in causing uterine infection (up to 75 per cent) and are usually indicated when a milky-white discharge is seen.

- *Staphylococcus*: *Staphylococcus aureus* is a widely distributed bacterium and is present in the reproductive tracts of infected and non-infected mares. It is known as an opportunist organism and it will invade and destroy tissue under certain conditions and is frequently associated with endometritis.
- *Pseudomonas*: *Pseudomonas aeruginosa* causes a venereal disease and reproductive tract infection in mares after mating with an infected stallion. Pseudomonas in the stallion's semen may be associated with reduced conception rates. However, many normal stallions carry small numbers of Pseudomonas organisms and show no reduction in fertility. Clinical signs of infection vary from profuse cloudy discharge that has specks of pus in it to no discharge at all, making diagnosis difficult. Control of the disease depends on early diagnosis and not letting the affected stallion cover mares naturally; here artificial insemination may prove useful in controlling venereal disease.
- *Klebsiella*: This causes venereal disease resulting in severe endometritis. Clinical signs vary in degree from none to a profuse mucopurulent discharge two to seven days after infection, and irregular oestrous cycles. Although the disease is usually transmitted during covering, mares can also be infected at teasing or during routine veterinary examination if adequate hygiene is not provided. Identification of venereal-producing strains of *Klebsiella* can be very difficult and in some cases can only be positively identified by test matings, i.e. if the organism is transmitted from mare to stallion to mare; however, this may not be practical. Treatment with antibiotics is sometimes extremely difficult as the organism appears to be insensitive to antibiotics. Again, the control relies upon rapid diagnosis and ceasing to allow stallions to cover naturally.
- *Haemophilus equigenitalis*: The organism causes the notorious disease contagious equine metritis (CEM) which caused disease of epidemic proportions in 1977. The Horse Race Betting Levy Board (UK) set up a committee of enquiry to control the disease, resulting in a code of practice being published (see Appendix) involving careful swabbing by a correct technique and examination by an approved laboratory. The stallion does not appear to be affected by the organism but he is a carrier transmitting it from mare to mare. *H. equigenitalis* causes acute or subacute infection; acute CEM leads to inflammation of uterine, cervical and vaginal membranes and a profuse grey discharge, three to five days after exposure. The mare may act as a carrier, infecting healthy

stallions without showing any symptoms herself. Due to the strict code of practice the occurrence of CEM in the UK is very low. However, the occasional acute case may still arise but greater awareness, better hygiene and stopping natural covering immediately should make control of CEM relatively simple.

- *Escherichia*: *E. coli* is commonly found in the intestine and can be transmitted by flies or droppings. It is the second most common organism isolated from the mare's uterus and is believed to be an important cause of uterine infection.
- *Equine herpes virus*: Equine coital exanthema is a specific herpes virus infection resulting in the formation of ulcers on the mare's vulva and stallion's penis. The infection is localised and has no effect on fertility but covering must stop. Mares usually recover in 10−20 days but may be left with white spots on the vulva.

In conclusion, there are many fertility problems that may affect the mare. However, many of these are relatively uncommon yet both the stallion and mare owner need to be aware of them in order to optimise fertility.

Abortion in mares

Abortion is the expulsion of the foetus before it is old enough to survive. There are many causes:

 (i) Virus
(a) Equine herpes virus 1 (Equine rhinopneumonitis)
(b) Equine viral arteriosus
 (ii) Bacteria
(a) Salmonella abortus equi
(b) Brucella abortus
(iii) Non-infectious causes
(a) Twin foals
(b) Stress/shook
(c) Failure of hormonal control
(d) Genetic defects

(i) Viral abortion

(a) Equine herpes virus 1 (Equine rhinopneumonitis):
This (EHV) is the virus generally associated with the abortion of

foals during the second half of pregnancy. The mare may, however, go full term but the foal dies shortly after birth.

Should a mare abort at this stage for no apparent reason then the foetus should be taken to the nearest Ministry vet for post mortem examination, and the mare isolated until the results have come back. This may only be a formality as the foal may have been aborted for another reason. If a mare aborts from EHV and loses her foal she will build up her own natural immunity, but this is short-lived, lasting only two to three months. The virus spreads rapidly, thus causing abortion throughout the whole herd. Isolation precautions are therefore very important as a vaccination programme is not always effective.

The precautions that should be taken on a stud where other pregnant mares are kept are:

- The mare that has aborted should be confined to her own loosebox or in one which is isolated from other pregnant mares.
- All areas which have been contaminated by the foetal fluids should be disinfected.
- All pregnant mares that have been in contact with the aborting mare for the previous three weeks should be isolated as a group as they may be infected.
- All other mares should be isolated as a potentially non-infected group and all traffic of horses on and off the stud should stop.

Once the cause of abortion has been diagnosed as not being EHV then these restrictions can be lifted. These immediate measures should be taken in every case of abortion unless there is clear indication that it was not a viral abortion. The Thorough-bred Breeders Association have a clear and detailed code of practice to help prevent and control EHV on which these guide-lines are based.

(b) Equine viral arteriosus (EVA)

This is another form of viral abortion, but unlike EHV there are obvious signs of illness such as filled legs, and discharge from the eyes and nose. When importing horses from the continent a negative EVA blood test result is required before entry is allowed into the UK. The chances of a positive EVA result are between 30–50 per cent. Further blood tests are required several months after the first test to monitor recovery before horses can be shipped.

(ii) Bacterial abortion

Salmonella abortus equi; *Brucella abortus*; *Streptococci*; *Klebsiella*; *Staphylococci*. These bacterial infections are common causes of abortion which can happen at any stage of pregnancy. Bacterial infections can enter the uterus through the cervix at any time before or after conception and the fertilised egg will not be able to survive in these infected conditions.

(iii) Non-infectious causes

(a) Twin foals

With modern techniques, which include the ultra-sound scanner, twin conceptions can be detected in the first 15−20 days of pregnancy and one of the twins pinched out. However, if this technique is not used, the twin pregnancy will be allowed to continue. At a later stage (approximately 4 months) one of the foetuses may die; it will start to decompose inside the uterus which will cause the eventual abortion of both twins. A small percentage of twins may be carried to term and survive but often one or both live foals are abnormally small and die soon after birth.

(b) Stress/shock

Although it is commonly believed that abortion can be caused by stress such as excess travelling, excitement causing mares to gallop uncontrollably when turned out in the paddock, and palpation by the vet when testing for pregnancy, these factors rarely cause abortion. By chance it may be that a foal is aborted soon after the local hunt has passed, but there is no proof that this is the certain cause of loss. However, excessive trauma can cause abortion, e.g. serious injury should a mare crash through a fence, a gruesome road accident, excessive pain from a violent colic attack, or even poisoning.

Although there is no real proof that abortion can be caused by these actions, it is best, by good stud management, to provide the mare with the most relaxed and natural gestation period as possible.

(c) Failure of hormonal control

Should a mare abort and there is no evidence of infection or any other reason found then it is likely that she has lost her foal due to hormonal reasons. The most common hormonal problem is failure of the mare to produce enough progesterone. (In a normal pregnancy it is this hormone which maintains the pregnancy.) If this is suspected, then regular blood tests for progesterone level should be taken. Should the level be lower than the required amount then expensive injections or oral administration of progesterone should be given. Even with this treatment there is still no guarantee that the mare will go full term. In some cases no reason can be found to prove the cause of abortion.

(d) Genetic defects
Often mares abort a malformed foetus and it has been suggested that the abortion of a defective foetus may be nature's way of controlling the genetic health of the species. Although there is little data available it appears that the older the mare is the more likely she is to abort a malformed foetus.

The stallion

The term 'infertility' is a broad one. Throughout this section the term *infertility* is used to describe a reduced ability of the stallion to impregnate the mare; *subfertility* indicates slight infertility, and *sterility* is the total inability of the stallion to fertilise the mare. If a stallion is *impotent* he has a reduced or complete inability to cover mares. Causes of stallion infertility can fall into three main categories:

  (i) Management
 (ii) Physical trauma and abnormalities
(iii) Reproductive disease

(i) Management

In the wild a stallion teases and covers mares at his own discretion; modern systems rely on in-hand mating, the time of mating being determined by teasing and the veterinary surgeon. Under this system the management of the stallion becomes of vital importance.

Nutrition
Malnutrition will adversely affect sperm production and sex drive. Feeding of the stallion is discussed in Chapter 8.

Over-use

The number of sperm per ejaculate depends on how quickly the testes can make sperm and the storage capacity of the epididymis. Over-use of the stallion results in fewer sperm per ejaculate and a greater number of immature sperm per ejaculate: thus stallion over-use results in temporary infertility and impotency.

Breeding season

Like the mare, the stallion is required to breed outside the normal breeding season. Artificially increasing day length during the winter or early spring will improve semen quality and libido.

Failure to recognise ejaculation

Some stallions show obvious signs of ejaculation, i.e. tail flagging, while others are not so obvious. Failure to recognise whether or not a stallion has ejaculated can be a substantial cause of infertility.

Painful covering

If a stallion has experienced pain while covering a mare he may be reluctant to cover, fail to ejaculate or dismount before ejaculation. Pain may be due to neglected feet, muscular pain, rough handling, a covering accident or being forced to cover a mare. Tact and patience are particularly important when training the young stallion to cover in-hand so that these circumstances can be avoided.

Age

Young stallions have limited sperm-producing capacity and should only have a small book of mares. Two year olds should only cover four to six mares, three year olds a maximum of 20 and four and five year olds 20 to 50. These figures are only approximate and will depend on the breed of horse, his career and whether there is a large veterinary input. As stallions age their sperm-producing capacity decreases and should be monitored and their book of mares reduced to compensate for the reduction in fertility.

(ii) Physical trauma and abnormalities

Cryptorchidism (rigs)

During gestation the testes migrate from the abdomen through the inguinal canal into the scrotum where the correct temperature for sperm production can be maintained. Retention within the abdomen of one or both testes is called cryptorchidism. If only one testis is retained (monorchidism) fertility may not be reduced but the retained testis may secrete extra testosterone causing increased male sexual behaviour and irritability. Cryptorchidism is an inheritable condition and breeding from such horses may spread the abnormality among the horse population.

Scrotal hernia
A scrotal hernia occurs when a loop of intestine passes through the inguinal canal into the scrotal sac. These are relatively common in newborn colts as the inguinal ring may not close for several months after birth. Scrotal hernia in the adult stallion is usually caused by trauma and possibly an inherited tendency to the disorder. The stallion will exhibit abdominal pain and altered gait and veterinary help will be needed to reposition the intestine via rectal palpation or surgery. Hernia may result in strangulation of the intestine which requires immediate attention if it is not to be fatal and also causes increased testicular temperature and hence reduced sperm production.

Physical trauma and injury
The penis, prepuce and testes are very vulnerable to injury resulting in psychological damage and possibly serious physical problems. If the penis is kicked when erect it will swell rapidly due to bleeding within the penis and should be treated immediately. A stallion may fall off a mare and injure himself, leading to an inability or reluctance to cover a mare.

(iii) Reproductive disease

Equine coital exanthema
As in the mare, this is a specific venereal disease caused by a herpes virus. The infection is usually brought in to the stud by a carrier mare that does not show symptoms; she infects the stallion and he subsequently infects other mares before clinical signs are observed and covering is stopped. The incubation period varies from five to ten days and takes about 10−14 days to clear up. Clinical signs are lesions on the penis, dullness and possibly fever.

Bacterial infection
The sheath and smegma contain many organisms, some of which may cause disease. These bacteria rarely affect the stallion and

their effect on mare fertility is more important. The bacteria are classified according to their effect on the mare:

(a) Organisms not associated with infection and endometritis in the mare, e.g. diphtheroids, *Neisseria* species (spp), *Streptococcus faecalis* and *Staphylococcus albus*.

(b) Organisms capable of causing infection in mares which have lowered natural defence mechanisms, e.g. *Strep. zooepidemicus*, *Staph. aureus*, *E. coli*, *Proteus* spp and some *Klebsiella aerogenes* types.

(c) Organisms capable of causing infection and venereal disease in mares with normal defence mechanism, e.g. some *Klebsiella*, some *Pseudomonas* and *Haemophilus equigenitalis* (the CEM organism).

The presence of these bacteria is detected by taking swabs from the urethra, urethral fossa, prepuce and pre-ejaculatory fluid.

Treatment is difficult as many of these organisms are very resistant to antibiotics and local treatment may remove natural bacteria allowing resistant bacteria to multiply.

Testicular disease
Any condition which results in a prolonged elevation of body temperature will have an adverse effect on fertility. Sperm production takes place at a lower temperature than that of the body and fever results in lowered sperm production. Due to time taken for sperm to mature prior to ejaculation the resulting drop in fertility may not be observed for some time after the fever has abated.

10 Artificial insemination and embryo transfer

ARTIFICIAL INSEMINATION

Artificial insemination (AI) is the technique used to inject semen into the mare's uterus using special insemination instruments. The use of AI has led to very rapid breed improvement, particularly in cattle; it is very rare these days for a dairy cow to be mated naturally. The technique has not taken off so rapidly in the horse industry and its use in the UK is still very limited although in the USA it is practical management in large studs. However, registration of stock conceived in this way is prohibited by some breed societies and it may not be practical on small studs due to the high capital input required to provide the special facilities, laboratory equipment and trained personnel needed. On the other hand the use of AI can overcome some fertility problems and increase conception rates if used properly.

Facilities, equipment and supplies

If AI is to be used extensively on a stud the facilities, equipment and supplies can represent a significant cost. However, a custom-built set-up is not essential and existing buildings and equipment can be used.

Covering area

This should meet all the criteria previously discussed; good clean footing, safe fencing or walls, sturdy wide gate or doorway, good lighting, etc.

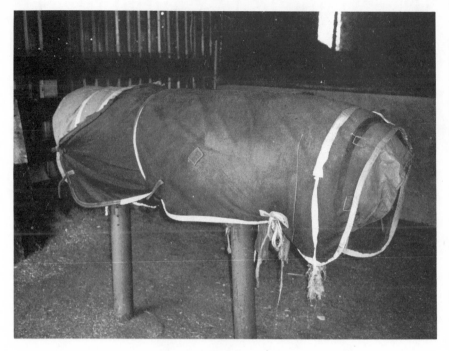

Fig. 10.1 Phantom mare.

Dummy or phantom mare

If the stud manager plans to use a phantom mare or dummy instead of a mare in season for the stallion to jump on it should be situated in the covering area (Fig. 10.1). The phantom should be the same size or slightly smaller than the stallion, padded and covered with a tough, washable material. To help the stallion maintain his grip during 'covering' the 'neck' of the phantom should be covered with material that the stallion can grip with his teeth.

Laboratory

Strict temperature control and careful semen handling are essential and a well-planned laboratory facilitates this. The lab should have good lighting and adequate power supply. Lab equipment should include incubator, waterbath, fridge, hot water, artificial vagina and insemination equipment.

Mare stocks

These are crates which restrain the mare so that rectal palpation, scanning, insemination and embryo transfer can be conducted safely. They should be of strong construction and made so that risk to the vet and mare are minimised.

All areas should be laid out in such a way that they can be thoroughly cleaned as hygiene is essential to success.

Semen collection

It must always be remembered that anything that makes contact with the genitals of mare and stallion should be as clean as possible and that anything touching the semen sample must be sterile.

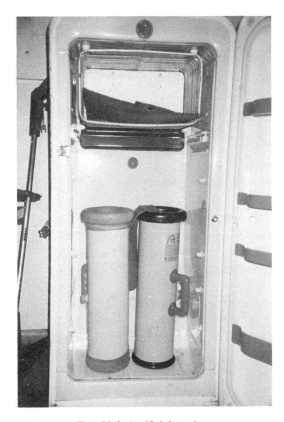

Fig. 10.2 Artificial vagina.

The artificial vagina

Semen is collected in an artificial vagina (AV). This consists of a
rigid container lined with two plastic or rubber liners (Fig. 10.2).
Warm water is introduced between the liners at a temperature of
49°–60°C (120°–140°F). This high temperature allows for cooling.
The water also controls the pressure within the AV so that when
the stallion's penis enters the AV it represents the mare's reproductive tract as closely as possible. The amount of water placed in
the jacket can be adjusted to fit individual stallions' preferences. If
the weight of the AV is noted the same pressure can be obtained
for future collections.

The inside of the AV is lined with a sterile lubricated obstetrical
sleeve or just lubricated with obstetric gel. A thermometer is left
inside the sleeve until the water between the liner cools to 44°–
48°C (110°–118°F). This allows enough leeway for the AV temperature to drop to body temperature between the lab and covering
area. A prewarmed collection bottle is attached to the end of the
AV with a filter to separate out the gel fraction from the semen
sample. Finally a prewarmed 'sleeping bag' is placed over the
collection bottle to insulate the semen from cold, shock and sun-
light (Fig. 10.3).

The collection procedure

It is extremely important to be well organised during the semen
collection. If an in-season mare is used the procedure is similar to
normal covering; if a phantom mare is used a mare in season may
be needed to make the stallion draw.

The mare handler, stallion handler and collector should all stand
on the mare's left side so that in an emergency the mare can be
pulled to the left so that her hindquarters swing away from the
stallion and both he and his handler are less likely to be kicked.
The stallion handler is also in a position to steady the stallion's
near foreleg so that it does not strike the collector.

The collector deflects the stallion's penis with a gloved hand
into the AV. The AV must not be forced on as this may discourage
the stallion (Fig. 10.4). Once the stallion has entered the AV it
should be steadied against the mare's thigh at the point of her
buttock with the collection bottle end pointed slightly down. If the

Fig. 10.3 Prewarmed 'sleeping bag' placed over collection bottle.

Fig. 10.4 The stallion's penis is deflected into artificial vagina.

stallion dismounts it is important not to follow him with the AV as this only reinforces any doubts he may have about the procedure.

After ejaculation, the stallion should be allowed to dismount in his own time. As he dismounts the AV should be removed and held so that the semen cannot run out and then taken to the lab as quickly as possible.

Semen storage and transport

The semen should be placed in an incubator at 38°C (100°F) if it is being used immediately or within 2 hours. Meanwhile a sample can be examined under the microscope and the sperm evaluated for fertility.

For transport, semen can be packed in a thermos-flask type arrangement and gradually cooled where it will keep for up to 48 hours. As yet freezing techniques have not been very successful with horse semen. Preparations called extenders are often used; these contain anti-bacterial agents and nutrients that help prolong sperm survival.

Frozen semen

Freezing semen allows it to be kept until needed. This means that frozen semen can be transported world-wide and that semen 'banks' can be accumulated from exceptional sires for use even after the stallion is dead. Although this technique is fairly successful in cattle it is still in experimental stages with horses. Of stallions tested so far approximately one-third have semen that will freeze successfully, in one-third some of the sperm will die and in the remaining third all of the sperm will die. This means it is essential for a preliminary freezing test to be made on the semen of any stallion using this technique.

Semen is collected by using an AV and diluted with a special semen extender containing nutrients and antibiotics. It is placed in sterile containers, sealed and frozen in liquid nitrogen. It is kept frozen until needed and then thawed and inseminated as normal.

Insemination

All equipment used to inseminate the mare must be sterile, including plastic tubing and catheter, an insemination pipette and a syringe for injecting semen into the tubing.

The inseminator then inserts a gloved lubricated hand, holding the catheter in the palm, into the vagina. The state of the cervix is checked with a finger (it should be relaxed), the index finger is then inserted into the cervical opening and the catheter guided into the uterus. The syringe is attached to the tubing or catheter, the semen injected into the apparatus and the insemination unit removed.

Advantages and disadvantages of AI

Disadvantages

(1) AI can have a detrimental effect on conception rates if done badly or by inexperienced personnel. The procedure needs to be well organised and well practised.
(2) Many breed societies are at present very wary of introducing AI. This is because it is their job to safeguard breeding identity and parentage and it may be much more difficult if AI is introduced. At present foals bred by AI are prevented from entry into any Thoroughbred Stud Book.
(3) AI can be expensive to set up if modern facilities are not available.
(4) AI can, but should not, be used as a means to overcome inherited reproductive faults, e.g. temperament and conformation.
(5) There may be abuse or error with respect to parentage because semen may be handled several times before it is used to impregnate a mare. Veterinary authorisation, blood typing and accurate record keeping are all essential.
(6) AI is thought to be likely to lead to the over-use of young or unproven stallions without proper testing before use. There may also be a danger of concentrating on just a few stallions, so limiting genetic variation in the population and leading to higher levels of inbreeding and inherited faults.

Advantages

(1) AI would decrease the spread of sexually transmitted disease because there is no physical contact between mare and stallion and antibacterial agents in extenders kill pathogenic organisms.
(2) It is possible to avoid the risks of damage and injury to both mare and stallion. Mares with Caslick's operation can be inseminated without re-opening and closing the suture line.
(3) AI can allow more effective use of an older, more valuable, proven stallion as his mares need not be limited because semen can be divided to be inseminated into several mares.
(4) Stallions which have developed behavioural problems or have been injured racing and are not suitable for natural coverings can still be used.
(5) Correctly performed AI can increase conception rates by avoiding over-use of the stallion and insemination of the mare at the optimum time for conception.
(6) The fact that semen quality is regularly checked under a microscope may help detect fertility problems before they are manifested by mares not getting in foal.

Conclusion

Artificial insemination in horses is likely to become widely used in the UK but there are as yet few stud managers and veterinary surgeons who are very familiar with the technique. AI would make breeding foals from stallions kept abroad or from mares that are unable to travel a practical proposition and its use, if carefully managed, should be welcomed by the horse industry.

EMBRYO TRANSFER

The embryo transfer process is well established in sheep and cattle but has only recently become a practical proposition in horses. The technique involves the transfer of a fertilised egg from a donor mare into a recipient or surrogate mare's uterus. The donor mare is covered by the stallion and the egg flushed out of the uterus seven to eight days after ovulation. Collection and transfer of embryos can be performed non-surgically with a reasonable success rate, making embryo transfer a practical commercial procedure in horses.

Reasons for embryo transfer

Embryo transfer has sound practical applications: for mares that get in foal readily and yet are unable to carry their own foals to full term; for mares that have foaling problems; and, possibly the most important, so that mares that are successfully competing can breed a foal without interrupting their careers. Normally a top class mare would have to wait until she has retired before she is allowed to breed a foal of her own, and by that time she may well be too old to breed.

In 1988 at the Warwickshire College of Agriculture two filly foals by Dutch Night (× Dutch Courage) out of the international three day event mare Morag were born. The first was born from Philemon, a half-Irish Draught mare, the day Morag was competing at the Badminton three day event. The second, a full sister, was born to Mattie, another Irish Draught cross, a month later (Fig. 10.5).

Requirements of the surrogate or recipient mare

The recipient mare should be in good physical condition and, when breeding competition stock, of a reasonably 'roomy' type so as not

Fig. 10.5 Morag (far right) with Mattie (far left) and Philemon (centre) and their two embryo transfer foals.

to affect the development and eventual size of the foal. Careful veterinary examination should be made before she is selected to be a recipient mare. She should be swabbed and free from disease and have good ovarian activity and cycle regularly every three weeks. Ideally, she should be a good breeder herself but may be of inferior quality to the 'donor' mare. Mares that crib-bite or wind suck should be avoided as these vices may be passed on to the potentially valuable foal.

The donor mare

The donor mare must be of superior quality, preferably a top class competition mare who will be able to pass on her attributes to her foal, even though she is still competing. She also should be free from disease and show a normal oestrous cycle pattern.

The method

(a) Synchronisation

Initially both donor and recipient mares' oestrous cycles must be synchronised, i.e. ideally the recipient mare must ovulate within 24 hours of the donor mare's ovulation. Their reproductive tracts are therefore in the same state when the embryo is transferred so that the recipient mare can 'take over' the donor mare's pregnancy and continue to maintain it.

This is accomplished by the use of the hormone prostaglandin which is used to bring the mares in season. It is given to both mares with a three day interval: so, if the donor mare is injected on Sunday, the recipient mare is injected on Tuesday. Luteinising hormone is then used following daily palpations and blood tests to ensure ovulation in both mares when required.

Alternatively, mares may be synchronised by allowing the donor mare to follow her natural cycle and giving the recipient mare LH during her season to cause her to ovulate, and four to five days later PG to bring her into season two to four days later. This way, her cycle can be shortened when required to bring her into line with the donor mare.

Thorough and regular teasing of both mares to establish whether or not they are in season and receptive to the stallion is essential.

Throughout this time as a back-up to teasing the veterinary surgeon will perform blood tests and rectal palpations to ascertain the reproductive state of the mares. Blood tests should continue to be taken during the first weeks of the surrogate's pregnancy. If the reproductive tract of the surrogate mare is not correctly synchronised, progesterone levels may not rise; progesterone is essential for the maintenance of pregnancy and it may be necessary to give the surrogate oral progesterone. Progesterone therapy may be discontinued once the uterus has taken over progesterone production from the corpora lutea (yellow bodies).

(b) Collection of the embryo

The donor mare is covered as normal when she comes into season. The veterinary surgeon may palpate the ovaries in order to establish whether there is a 'ripe' follicle present before the mare is covered. Covering can be risky for both mares and stallions and the owners of the valuable competition mare may wish to minimise the number of times the mare has to be served. Ovulation of a ripe follicle can be encouraged after the mare has been covered by an injection of luteinising hormone. Seven days later the tiny embryo, just visible to the naked eye, is removed from the donor mare's uterus and transferred to the recipient mare; this may be done surgically or non-surgically. The non-surgical technique (although fertility rates are lower) is now most commonly used; this involves the washing or flushing of the embryo from the uterus.

The donor mare is put in specially designed 'stocks' which keep her still during the flushing, with an experienced handler at her head. Cleanliness is vital so the tail is carefully bandaged and secured out of the way. The vulva is then washed with mild disinfectant; this area is then washed with saline solution to remove any traces of disinfectant as this may be lethal to the embryo.

A catheter is then inserted into the vagina and through the cervix; an inflatable cuff blocks off the cervix and prevents the tube from falling out during flushing (Fig. 10.6). A specially prepared nutrient solution kept at body temperature is then passed into the uterus via the catheter. About one to three litres will be used to fill the uterus. The vet will palpate the uterus per rectum and once it

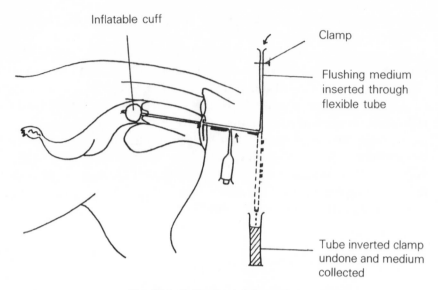

Inflatable cuff

Clamp

Flushing medium inserted through flexible tube

Tube inverted clamp undone and medium collected

Fig. 10.6 Collection of the embryo.

contains sufficient fluid and becomes turgid his assistant will release a valve on the catheter, allowing the solution to pour into sterile cylindrical flasks which are also kept at body temperature. As the uterus is emptying, the vet gently palpates the horns and body of the uterus to encourage the embryo to be flushed out with the fluid. If an embryo is not found then the process may be repeated. The collected fluid is allowed to settle and is then taken to a laboratory, preferably adjacent to the collecting area, and examined under a low power microscope, although an embryo may have already been observed, as it may be just visible, by the experienced embryo spotter! The flasks are allowed to stand for 20 minutes and the fluid is then pumped off to leave a small quantity containing the embryo which has sunk to the bottom of the flask. The embryo is then collected by sucking it up into a pipette and immediately transferring it to the surrogate mare.

(c) Transfer of the embryo

Transfer can be done either surgically or non-surgically. At present, surgical methods give a better conception rate but as non-surgical techniques become more sophisticated so their success rate improves.

Surgical method (Figs. 10.7 to 10.14)

The recipient mare is given a general anaesthetic and a small midline incision is made just in front of the udder. The vet is then able to bring out one horn of the uterus, pierce a small hole just big enough to insert the nozzle of the pipette, and the contents of the pipette are injected into the uterine horn and the mare is sewn back up. The operation is not very traumatic and the advantages of

Fig. 10.7 The donor mare's uterus is filled with a special medium.

this method are that the tightly closed cervix of the mare in dioestrus is not interfered with and the embryo is placed directly where it ought to be for that stage of pregnancy.

Alternatively, the mare can be placed in stocks and sedated and operated on under local anaesthetic, making an incision in her flank just in front of the hip bone, and then putting the embryo directly into the uterine horn.

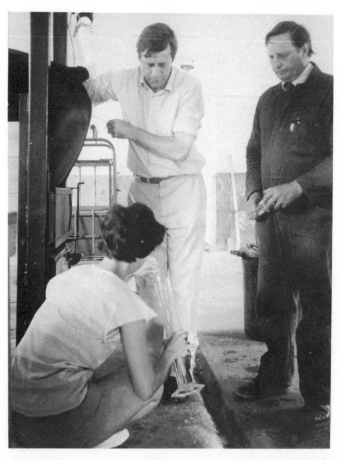

Fig. 10.8 The medium is allowed to pour from the uterus into collection flasks. The uterus is gently palpated to aid this.

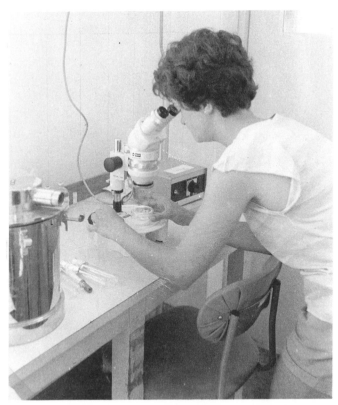

Fig. 10.9 The medium is searched for a fertilised egg.

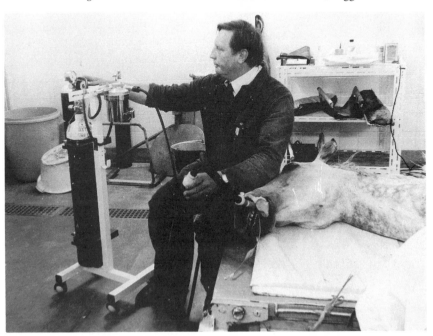

Fig. 10.10 The recipient mare is anaesthetised.

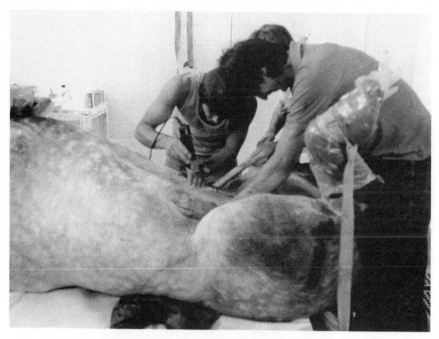

Fig. 10.11 The mare is prepared for surgery.

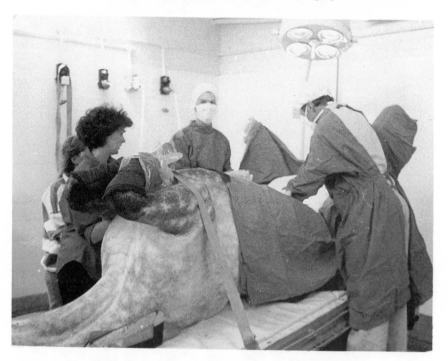

Fig. 10.12 An incision is made to expose the uterine horn.

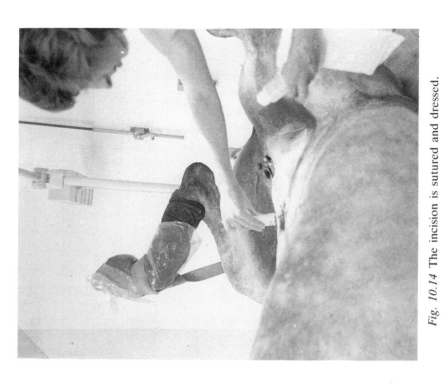

Fig. 10.14 The incision is sutured and dressed.

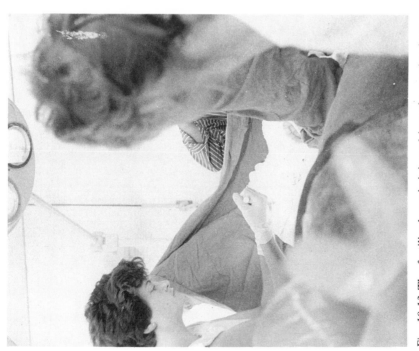

Fig. 10.13 The fertilised egg is injected into the uterine horn.

Non-surgical method (Figs. 10.15 to 10.25)

In this method the mare is put in the stocks and she is sedated if necessary. The embryo is loaded into a special inseminator which is then passed into the uterus by careful penetration of the cervix and a plunger on the end of the catheter is pressed which pushes the embryo directly into the uterus. This is less traumatic for the mare and less expensive for the mare owner but there is a greater risk of damage to the embryo and a possibility of introducing infection through the cervix. Antibiotics should therefore be administered directly after the transfer has taken place.

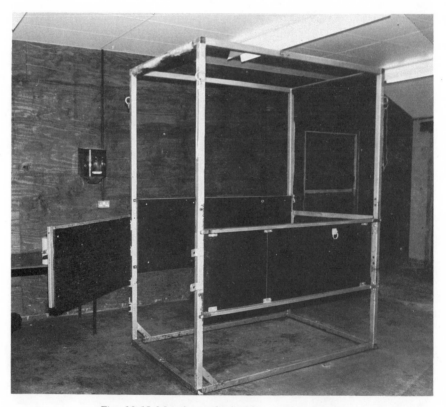

Fig. 10.15 Mare's stocks inside veterinary unit.

 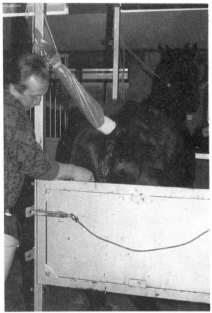

Fig. 10.16 The donor mare's tail is tied up and bandaged.

Fig. 10.17 The mare's perineal region is thoroughly cleaned.

Fig. 10.18 A catheter is inserted into the uterus.

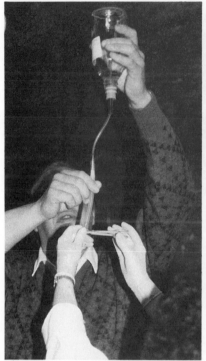

Fig. 10.19 The mare's uterus is filled with flushing medium.

Fig. 10.20 The medium is then released and collected in sterile cylinders.

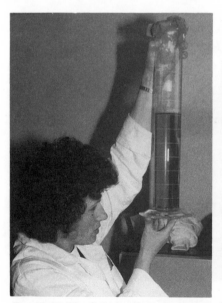

Fig. 10.21 The fluid is allowed to stand for 20 minutes and the majority is filtered off.

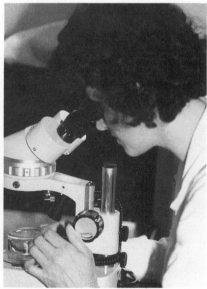

Fig. 10.22 Closer examination is made of the remaining fluid.

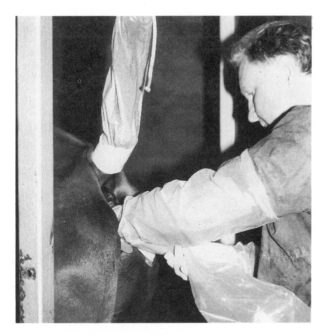

Fig. 10.23 A special catheter is inserted into the recipient mare's uterus.

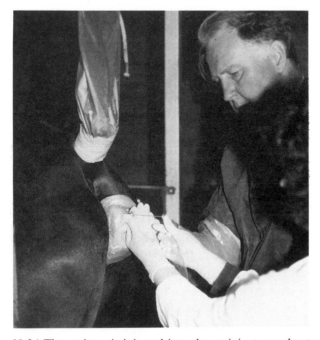

Fig. 10.24 The embryo is injected into the recipient mare's uterus.

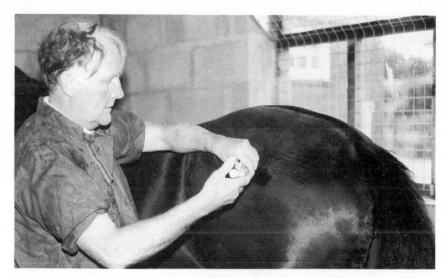

Fig. 10.25 Finally the recipient mare is given an antibiotic injection.

(d) After transfer

As previously mentioned it is wise to blood-test the recipient mare to monitor the progesterone levels; she should also be scanned for pregnancy approximately fourteen days after the transfer. If she is not pregnant the procedure can be restarted.

The donor mare should also be scanned as it is possible for her to have conceived twins − donated one and be pregnant with the other.

Once pregnancy is confirmed the recipient should be treated like any other broodmare and the donor may return to work. It is, however, an advantage to 'let down' the donor mare as much as possible as it is known and accepted that very fit mares are more difficult to get into foal.

This process can, of course, be repeated several times throughout a season. However, it must be noted that to do this would create a 'battery farming' image which must be avoided, added to which the value of top class stock will drop as better horses will become more easily available. It is for these reasons that only one or maybe two foals per year per donor mare should be transferred. It must also be pointed out that societies which accept ET methods generally will only register one foal per year, and the thoroughbred industry at present refuses to register any stock at all.

11   Parasite treatment and control

A parasite is an animal which lives on or in another, called the host, at the host's expense. Parasites fall into two categories: internal which live inside the host, and external which live on the exterior of the host. External parasites (ectoparasites) which affect horses are ticks, mites and lice. The major internal parasites (endoparasites) are shown in Table 11.1.

Internal parasites

Commonly known as 'worms' internal parasites are a well-recognised cause of ill-health in adult and young horses. Worm control is a vital aspect of stud management and can be made more effective by an understanding of the life-cycles of these parasites and a knowledge of the drugs available for their control.

Horses are susceptible to different types of worm at different stages in their life (see Table 11.2). Of primary importance in the young foal are ascarids and threadworms while the large redworm (*Strongylus vulgaris*) is by far the most serious worm infection in older horses.

Ascarids

These are large worms which may be more than 12 inches (30 cm) long and as thick as a pencil. An adult egg-laying female living in the small intestine can lay up to 100,000 eggs per day. These eggs have a tough sticky outer coat which makes them very resistant to disinfectants and the environment allowing them to survive for years outside the horse. Larvae develop within the eggs and when the eggs are eaten by a foal the larvae hatch in the foal's gut and then burrow through the gut wall and migrate through the liver to

Table 11.1 The major internal parasites of horses

Type	Species	Development site	
		Adults	Larvae
Large strongyles (redworm)	Strongylus vulgaris	Caecum	Intestinal arteries
	S. edentatus	Colon	Liver
	Triodontophorus spp.	Caecum/colon	Intestinal wall
Small strongyles	Trichonema spp.	Caecum/colon	Intestinal wall
Roundworms (ascarids)	Parascaris equorum	Small intestine	Liver/lungs
Bots	Gasterophilus spp.	Bot flies	Stomach
Threadworms	Strongyloides westeri	Small intestines	Various tissues
Pinworms	Oxyuris equi	Colon/rectum	Intestinal wall
Lungworms	Dictyocaulus arnfieldi	Bronchi of lungs	Lymphatics and lungs
Tapeworms	Anoplocephala perfoliata	Ileum/caecum	(Intermediate host)

Table 11.2 Types of worm affecting different ages of horses.

Foal	Threadworm	(*Strongyloides westeri*)
	Large roundworm White worm	}(Ascarid)
	Large strongyles Large redworm	}(*Strongylus vulgaris*)
Up to three years	Large roundworm Large strongyles Small strongyles Bots Tapeworm Lungworm Pinworm (*Oxyuris equi*)	
Adult	Large strongyles Small strongyles Bots Tapeworm Lungworm Pinworm	

the lungs. They are then coughed up, swallowed and undergo final development to become egg-laying adults in the small intestine. It takes 12 weeks from the foal eating the infective eggs for the larvae to mature and for the foal to start passing out eggs in the faeces and thus infecting the pasture.

Foals may have up to 1000 adults in their gut, giving them a dull coat, a pot-belly and possible retarded growth. Adult worms may actually block the gut causing it to rupture leading to peritonitis and death. The migrating larval stages may cause coughing and nasal discharge.

Very heavy pasture contamination can occur on paddocks that are grazed every year by mares and foals. Mares should be wormed regularly during pregnancy and shortly before foaling to reduce the number of eggs in the environment. Stables must be clean and mucked out daily. Ideally foals should be turned out into 'clean pasture', i.e. not grazed by horses for at least 12 months, but in practice this is not usually possible and so foals must be wormed from four weeks of age and every four weeks thereafter with an effective wormer. This will help avoid serious contamination of pasture for next year's foals. After the yearling stage horses appear to develop resistance to ascarids and they do not cause ill-health in adults.

Threadworms (Strongyloides westeri)

This is the first parasite to which the foal may be exposed since larvae can be passed from mare to foal in her milk. Infection can also occur by larvae penetrating the foal's skin. Threadworms, as their name suggests, are very small and generally well tolerated, but heavy infection can cause diarrhoea (scouring) which may coincide with the mare's foaling heat. Scouring foals can dehydrate very easily and any scouring that does not stop after the foaling heat should be treated immediately. Foals may be routinely treated for threadworm from seven days old, though some wormers need to be given at a higher dose than normal in order to be effective. Regular worming of mares will help reduce the incidence and severity of threadworm.

Strongyles

This is the most important group of parasites, found in horses of all ages, and living in the large intestine. These worms are divided into two groups depending upon their size but their life cycles and effect on the horse are similar. Eggs are laid by adult worms in the gut and, passed out in the dung, larvae hatch and develop on the grass and become infective third stage (L3) larvae. The speed of this development depends upon climate and proceeds most rapidly in warm, wet conditions. The L3 larvae are eaten by the grazing horse and pass into the intestines. Here the life cycle varies between species; small strongyles burrow into the intestine wall where they develop to re-emerge three months later as egg-laying adults. *Strongylus vulgaris* (large redworm) migrates extensively through the body; about eight days after infection the larvae develop to fourth stage larvae (L4) which migrate to the anterior mesenteric artery, which is responsible for supplying most of the gut with blood. This migration damages the walls of the artery and may lead to blood clot formation. These clots may block smaller blood vessels and disrupt blood supply to the gut causing colic (redworm are thought to be the commonest cause of recurrent bouts of spasmodic colic). Eventually mature larvae return to the large intestine where they become egg-laying adults approximately 200 days after infection.

This prolonged migration may lead to 'false negative' faecal worm egg counts; there may be no adult egg-laying worms in the gut and thus a negative worm-egg count will be recorded, even though the tissues may contain many larvae which are causing extensive damage and will eventually become adults in the gut. Both large and small strongyles can remain dormant in the gut wall so that when egg-laying adults are removed by worming a new wave of larvae emerge and become adults. Consequently, if the horse has been heavily infected repeated routine treatment may not solve the problem. Few wormers are effective against these migrating stages but providing horses are regularly wormed so that pasture contamination is low the problem will be minimised. Some wormers are effective against larval stages in large repeated doses and when horses are brought up from grass it may be useful to give a larvicidal dose to reduce the population of developing strongyles.

Pinworm (seatworm, Oxyuris equi)

This worm is killed by routine doses of wormer and thus is not generally of significance. The adults live in the large intestine and the female lays her eggs on the skin surrounding the anus which causes intense itching. Larvae develop within the eggs in four to five days and the eggs then fall on to the pasture where the grazing horse eats them. After ingestion the larvae hatch from the eggs and develop in the wall of the large intestine before completing their development in the lumen of the intestine.

Bots

Horse bot flies lay their eggs on the hairs of the legs, shoulders and neck during the summer months. When the horse licks itself it takes eggs into its mouth which travel to the stomach where they develop and mature before passing out in the faeces the following Spring. After pupating the adult flies emerge to complete the cycle. Traditionally horses have been treated for bots after the first frosts, i.e. November and again in the Spring to remove any bots

that did not pass out spontaneously. The extent of the damage that
the presence of bot maggots in the stomach causes is not completely
understood but large populations can cause ulceration.

Lungworm

Lungworm can cause acute coughing but even badly affected horses
may not pass out larvae in the faeces. This contrasts with the
donkey where coughing is rare and yet larvae are commonly found
in the faeces. Diagnosis of lungworm in horses is therefore difficult
and is often made only after successful treatment with an effective
wormer.

Tapeworm

The adult tapeworm is found in the caecum, particularly where the
small intestine enters the caecum (ileocaecal junction). The adult is
8 cm long and 8–14 mm wide and attaches onto the gut wall by
suckers; it sheds egg-containing segments which are passed out in
the dung. The eggs are eaten by an intermediate host, oribatid or
forage mites, where development takes place. The horse is infected
by eating the mites. and once ingested it takes six to ten weeks for
the horse to start passing out eggs in the faeces. Tapeworm infes-
tations have been implicated in colics, peritonitis and digestive
upsets. It is difficult to diagnose tape worm infection by faecal egg
counts and many more horses may be affected than previously
thought. Few wormers are effective against tapeworm but it may
be advisable to treat foals and adult horses (excluding pregnant
mares and stallions) twice a year with an effective wormer.

Treatment and control

Tables 11.3 and 11.4 show some drugs currently marketed for use
against internal parasites of the horse, and which drugs are effective
against the different types of worm. Correct and efficient use of
these drugs is a vital part of stud and mare management; most
studs have limited access to grass and the same paddocks will be
used for mares and foals year after year. During the height of the

Table 11.3 Drugs available for control of internal parasites.

Trade name	Supplier	Active ingredient	
			Benzimidazoles
Pony and Foal Wormer	Crown Chemical Co. Ltd	Piperazine	
Equizole	Merck, Sharp and Dohmé	Thiabendazole	⎫
Thiabenzole	Merck, Sharp and Dohmé	Thiabendazole	⎪
Equivurm-Plus	Crown Chemical Co. Ltd	Mebendazole	⎪
Telmin	Crown Chemical Co. Ltd	Mebendazole	⎪
Panacur	Hoechst UK Ltd	Fenbendazole	⎬
Synanthic	Syntex Agribusiness	Oxfendazole	⎪
Systamex	Wellcome Foundation	Oxfendazole	⎪
Bayverm LV	Bayer UK Ltd	Febantel	⎪
Equitac	Smith Kline Animal Health	Oxibendazole	⎪
Rycovet	Rycovet Ltd	Oxibendazole	⎭
Strongid-P	Pfizer Ltd	Pyrantel Embonate	
Atrobot	Arnolds Veterinary Products Ltd	Dichlorvos	
Neguvon	Bayer UK Ltd	Metriphonate	
Eqvalan	Bayer UK Ltd	Ivermectin	

Table 11.4 What wormer to use when.

Parasite	Wormer	Comment
Routine worming	Equizole	
Adult strongyles	Telmin	
	Panacur	
Oxyuris equi (pinworm)	Systemex	
	Bayverm	
	Equitac	Effective against strains resistant to other benzimidazoles
	Strongid-P	
	Eqvalan	Does not kill eggs
Strongyloides westeri in foals: 1–4 weeks	Eqvalan	
	Equizole	
	Equitac	
	Panacur	Requires high dose: 25 ml Panacur 10% or half syringe
Ascarids in foals: from 4 weeks to 8 months dose every 4 weeks	Equivurm/Telmin	
	Equitac	
	Systamex	
	Panacur/Equizole	High dose rate
	Strongid-P	
	Eqvalan	
Strongyle larvae:	Eqvalan	
	Systamex	
	Panacur	Dose rate daily for 5 days
Lungworm	Eqvalan	
	Panacur	High dose rate (2×)
	Telmin/Equivurm	High dose daily for 5 days
Bots	Neguvon	
	Eqvalan	
	Frizk	
	Astrobot	
Tapeworm	Strongid-P	Dose × 2

stud season large numbers of visiting mares may be turned out together, some of which may not have been wormed regularly. The stud policy must be to worm every four to six weeks in order to suppress worm egg output and thus reduce the parasite population on the pasture to low levels. Any new arrivals should be treated immediately on arrival and then turned out with the other horses.

In any parasite control programme, there should be a change over to a chemically unrelated compound every 12 months in order to avoid parasites building up a resistance to one wormer. Small strongyles have been shown to be able to develop resistance to wormers containing benzimidazoles (see Table 11.3).

Good grassland management will help control the free-living stage of the parasite. Appropriate techniques include ploughing and reseeding, rest and rotation, mixed grazing, removal of droppings, stocking rates, harrowing and topping.

Ploughing and reseeding

This may be the only remedy for really horse-sick pasture. Ploughing to a depth of 9 inches (23 cm) should put eggs and larvae where they cannot survive although some sward may still be exposed. The paddock will be out of use for about 12 months and it is only viable where grazing is not very limited.

Rest and rotation

Paddock rotation which allows mares and foals to avoid grazing fields used the previous year by horses is very useful but rarely practical where grazing is limited. Any break in grazing will help reduce the worm egg population of a paddock, e.g. taking silage or hay crop. It would be advisable to avoid feeding this hay to mares and foals.

Mixed grazing

Equine parasites do not affect sheep and cattle, which will eat and destroy many infective larvae. Cattle will also eat areas of grass that horses leave to get long and rank.

Removal of droppings

Whilst being expensive in terms of labour, picking up the droppings at least twice a week improves appearance and avoids areas of rank growth. However, eggs and larvae will still be left behind to some extent.

Stocking rate

The fewer horses on a paddock the smaller the parasite problem, but keeping well stocked is a necessity during the high stud season. Similar classes of stock should be grazed together so that they can be wormed in accordance with their needs, which may vary with age.

Harrowing and topping

These practices will help prevent selective grazing and keep pasture in good condition. Harrowing should only be carried out during warm dry weather so that larvae are killed by exposure to the drying effect of the sun. Spreading the droppings during wet weather merely serves to spread the larvae over the whole paddock.

External parasites

Lice

Lice are the most common external parasites of horses. They fall into two categories: the sucking louse and the biting louse. Lice are wingless insects which lay their eggs or nits on the horse's body, glued to the hair, a favourite area being the roots of the mane or tail. They cause intense itching so that the horse rubs itself, losing hair and developing raw patches. The horse may also lose condition partly because the lice are blood-sucking and partly due to the time spent scratching and not eating. Lice are generally associated with poor or neglected horses but out-wintered broodmares with long coats are very susceptible and the lice are highly contagious.

There are several powders and washes available which kill the adult lice. However, the eggs are resistant and applications may have to be repeated several times to kill these eggs as they hatch.

Ticks

Ticks are blood-sucking parasites which affect all animals. They attach themselves to the host, gorge themselves for several days and then drop to the ground. In many parts of the world ticks are responsible for the transmission of serious blood-borne diseases. The ticks should either be allowed to feed and then drop off or killed by treating the horse with Eqvalan. Under no circumstances should the tick's body be pulled off as the mouth parts would be left behind and may cause abcess formation.

Mites

Mites are small insects that cause mange. This disease falls into three categories: sarcoptic, psoroptic and symbiotic mange. The first two are notifiable diseases and are all but eradicated from the UK. Symbiotic mange is usually confined to the legs or root of the tail, more commonly in the heavier breeds with heel feathers. It causes great itching and stamping. The veterinary surgeon will advise on treatment.

Ringworm

Ringworm is a contagious skin disease caused by fungi which is easily transmitted via tack, grooming kit, stables and contact with affected animals. Initially, the hair stands up in small circular patches which may grow quite large. The hair then falls out leaving bald patches which may become crusty. There is usually little or no itchiness but in badly neglected cases where horses are kept in dirty conditions the horse may lose a lot of condition.

Ideally, affected horses should be isolated, the hair round the lesions removed and burnt and the scabs treated with fungicidal washes and sprays. The veterinary surgeon can prescribe an oral treatment but it may not be advisable to give this to infoal mares.

If left to run its course, as in the wild, horses will eventually recover from ringworm and develop a certain amount of resistance to re-infection, and some yards choose to adopt this strategy. Whilst ringworm is unsightly and requires prompt treatment it is rarely serious in the long term.

12 Stud control

This chapter deals with some management aspects of running a stud, including buildings and facilities, stud records and grassland management.

Stud design

The ideal stud has been carefully designed and built from scratch and subsequently soundly managed as a business so that mares, stallions and young stock can be safely and efficiently cared for. The following chapter outlines the basic requirements for a two stallion stud.

The setting

Well-drained fertile soil is a great asset for a successful stud enterprise; horses will be able to stay out for longer in the winter without poaching the ground and good grass growth will help minimise feed bills. It is not always possible to get the ideal site and provision must then be made for bringing horses in earlier and grass growth starting later. Once a suitable site has been found it must be carefully maintained so that pasture quality does not decline.

Stallion boxes

Horses require roomy, well ventilated boxes and the stallion is no exception: he requires a box at least 14 feet by 14 feet (4.27 m × 4.27 m), and preferably bigger to prevent boredom. Traditionally, stallions have been kept in separate yards, well away from the mares' yard but it may be desirable for stallions not to be completely isolated from the yard activities. If stallions are being frequently visited on a public stud it is wise to have strong grids on the doors to prevent the curious onlooker from constantly harassing the stallion. The stallion's name should be prominently displayed and the box should always look clean and tidy when prospective clients arrive.

Mares' boxes

There should be at least 30 boxes for visiting mares; mares with a foal at foot require boxes of at least 12 × 14 feet (3.66 m × 4.27 m), and preferably 14 × 14 feet (4.27 m × 4.27 m). A barn system is economical in terms of space and labour and can usefully take advantage of existing buildings. Care must be taken to ensure that there is adequate ventilation.

Ventilation

Ventilation replaces stale air with fresh air and thus reduces high humidity and condensation, both factors which contribute to respiratory problems in horses. Ventilation can be provided by using windows, vents or fans. Vents and the open part of the window should be at least six feet (1.8 m) from the floor so that the air moves freely through the box without causing draughts at eye level. Fans can be used in large barns to force air flow; these help cool the air in summer and prevent stuffiness in winter.

Flooring

Drainage is an important consideration in stable design. The floor should be sloped evenly at a rate of about 1 in 60 (1 inch in 5 feet) and should end in a gutter running to the main drain.

The flooring should be durable, resilient and offer secure footing. Concrete and brick can be scrubbed and disinfected without damaging the surface but can become slippery when wet. This means that a good thick bed is essential. Bedding should be absorbent, free from dust and economical; good quality straw is ideal, particularly for foaling boxes. Wood shavings can be used but tend to stick to the nostrils of a wet newborn foal.

Lighting

Stables should be designed so that as much natural light enters as possible, but an efficient artificial light source is essential. Lights should illuminate all indoor facilities and outside approaches; outside security lights are also a good idea. Lights should be covered by mesh and placed so that they cannot be reached by horses. Switches should also be protected and safely positioned.

Isolation boxes

One or two isolation boxes, situated well away from the yard, are essential to stop the spread of infectious disease. They can be used to house mares on arrival at the stud if it is suspected that they have been in contact with disease.

Foaling boxes

There should be one or two spacious foaling boxes; these should be at least 14 × 16 feet (4.27 m × 4.88 m) and have a minimum of stable fittings on which the foal might injure itself. Ideally, there should be heat lamps and closed circuit television so that foalings can be carefully monitored. The boxes should be in a quiet situation with a sitting-up room containing the television monitor nearby.

Youngstock yard

It may be advantageous to over-winter young horses in a yard rather than stable them individually. They can be put in their own stables at night and released into a covered yard by day to exercise and play. This provides them with a more healthy and natural environment. The yard must be free from dangerous projections and of an adequate size for the number of youngsters.

Teasing and covering yard

The teasing yard should be an enclosed, and preferably covered, area convenient to the stallion boxes containing a teasing board. The board should be no less than four feet (1.25 m) high and ten feet (3 m) long and made of a narrow pyramid of heavy wood supported by wooden or metal posts set in concrete. The board should be heavily padded with rubber to prevent injury to both mare and stallion and it is sometimes fitted with a roller bar along the top to help the stallion should he accidentally get a leg caught over the top of the board. The teasing and covering yard surface should be non-slip so that the stallion can get good footing when he covers a mare; it should also be dust-free.

Examination stocks

Stocks can be used to prepare mares for covering, pregnancy testing, artificial insemination and embryo transfer. They are designed along lines similar to cattle crushes, being sturdy cubicles used to restrain mares. Ideally, solid partitions should enclose the mare on two sides with two doors allowing access and exit (Figs. 12.1, 12.2, 12.3). Other buildings should include a tack room, feed room, office, wash box (hot and cold water and heat lamp), staff accommodation, garages for machinery, a loading ramp and hay barns.

The paddocks

About 75 acres (30 hectares) will be required for this stud with two stallions covering approximately 100 mares. Stock should be divided into appropriate groups, e.g. mares and foals, barren mares, foaling mares, stallions and youngstock. There should be one paddock resting for each group so that fresh grazing is available at all times.

Fencing

Double post and rail fencing about 4'6'' (1.37 m) high is ideal between paddocks although existing hedges, providing they are

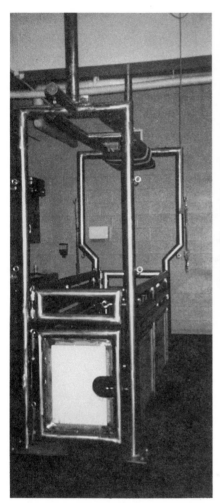

Figs 12.1 and *12.2* Mare's stocks

sturdy, are useful as they also provide shelter. Corners should be rounded so that horses cannot be trapped and bullied and additional rails put over any low gates onto roads. Wherever possible adjoining paddocks should be of equal length; occasionally injuries occur if two horses are running in neighbouring paddocks of unequal length.

It may be useful to include a teasing board into the fence line so that in-season mares can be detected in more natural surroundings. Whatever type of fencing is used, it must be sturdy, require low maintenance, and be highly visible and safe. Hazardous areas must

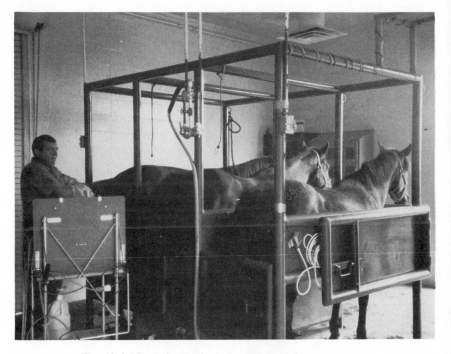

Fig. 12.3 Mares in stocks being scanned for pregnancy.

be fenced off and roadside fences must be particularly strong. Post and rail is visible, attractive to look at and safe, but it is expensive, requires constant maintenance and has a limited life span. Oak is strong and durable but has a tendency to warp and split into long spear-like splinters when broken. Pine is not so durable but does not splinter. Any wood used for fencing should be treated with preservative to extend its useful life.

Specially designed wire stud fencing appears to be a safe and cheaper alternative to post and rail. The mesh should be small enough to prevent horses getting their feet through and must be well strained with a visible band along the top (Fig. 12.4).

Another alternative is plain wire attached to wooden posts with a wooden rail or visible band along the top. High tensile wire may not be suitable as there is no 'give' in it should horses gallop into it. Barbed wire should never be used on stud farms.

Fig. 12.4 Wire stud fencing with visible band along top.

Shelter

Paddocks should have sheds to provide shade and shelter. Ideally, the inside corners of the shed should be rounded and support posts padded to prevent injury.

Water

Horses must always have a clean, fresh water supply, best provided by conveniently placed water troughs which are regularly drained and cleaned. Broodmares with foals at foot drink a tremendous amount of water on a hot summer's day and the tanks must be large and refill quickly to supply the mares' needs.

Stallion paddocks

Stallion paddocks must be strongly fenced and carefully situated so that the stallion does not become excited by having mares and foals or strange horses in constant view. The fence should be at least five to six feet high, and in some cases a strand of electric wire pulled taut over the top rail may lessen the risk of the stallion jumping out, rearing over the fence or chewing it.

Accessories

Space will also have to be given to car and lorry parking, the muck heap, and small, safe turn-out paddocks or 'playpens' for newly-born foals or youngstock.

Fire prevention

Stable fires are tragic and destructive, and when planning a stable yard it is vital to eliminate fire hazards as much as possible. Local fire officers are available for advice and will inspect a property for safety.

High risk areas such as hay and straw barns should be surrounded by a buffer zone of 100 feet (30 m) if possible. Buildings should be constructed of suitable fire-resistant materials. All wiring and electrical fixtures should be professionally installed and inspected routinely. Other fire safety precautions include wire cages surrounding light fittings, *NO SMOKING* signs, and regular sweeping and dusting of floors, walls and rafters to remove dust and cobwebs. Electrical appliances, particularly heaters, must be used with care and not left unattended surrounded by drying tail bandages. Fire walls constructed of brick and coated with fire-resistant substances can be built between storage and stable areas to slow down the spread of fire, and all doors should be closed at late check.

In the event of a fire there should be an effective, practical and practised fire drill. The drill must be clearly posted by the fire alarm and telephone. It is most important that the fire alarm is easy to find, audible and dependable.

Portable fire extinguishers should be placed in high risk areas such as feed rooms, storage areas and offices. Extinguishers are designed to handle specified types of fire and should be chosen accordingly, e.g. where a fire may be started by an electrical fault a blue chemical extinguisher should be used, whereas red water extinguishers would be suitable in feed rooms.

Fire hoses should be positioned so that all buildings can be reached. These hoses must not be for everyday use as faults may develop and not be discovered until it is too late.

The fire drill

(1) Sound the Alarm.
(2) Dial 999 and the Stud Manager.
(3) If fire is small, extinguish it.
(4) If fire is too large to contain, remove any horses in immediate danger to specified area and account for all staff.
(5) Shut doors and windows and protect nearby buildings by hosing them thoroughly.

Fire drills may seem inconvenient and a waste of time but they are useful in educating staff and finding any flaws in the emergency procedure.

Stud records

It is important to keep correct and strict records to ensure the efficient running of a busy stud. These should include mares' covering charts, nomination forms, details of daily veterinary visits, farrier's visits, worming records, covering certificates and registration, and invoicing.

Mares' covering records

This is probably one of the most important set of records made and should be kept for several years as a reference.

Centrally placed wall charts should be kept, preferably near to a telephone, so that mare owners can have an immediate update or progress report on their mare while she is away at stud. Even if the stud groom is not available, other members of staff should be trained to understand these charts and keep owners informed and happy. It is also essential that these charts are completed as soon as the teasing and covering has been finished, by means of a coding system. An example is shown in Fig. 12.5.

MARE	OWNER	MAY																					28	29	30	31
		1	2	3	4	5	6	7	8	9	10	11	12	13	14	15	16	17	18	19	20		28	29	30	31
STAR	Smith	O A	W	O		O		V TO	X		X		O											O	V PD+		D
DAISY	B.S	FF						V TO		TO														X TO	X TO	O	

Key: Arrived (A); Departed (D); Foaled Colt (FC); Foaled Filly (FF); Vet (V); Tried off (O); Tried On (ON); Covered (X); Pregnancy Diagnosis Positive (PD+); Pregnancy Diagnosis Negative (PD−); Wormed Mare (W); Wormed Foal (WF); Blacksmith (B).

Star arrived 1 May and was wormed on 2 May. She was tried every other day until she came into season (7 May). She was swabbed clear and covered on 8 and 10 May, and on 12 May she was no longer in season. Twenty-one days from the first day she was in season, she was tried and did not come back into season. The following day she was scanned in foal (19 days after her last service). On 31 May she went home.

Daisy foaled on 1 May and although she was swabbed clear on her foaling heat she was not covered until her subsequent season.

Fig. 12.5 Section of a mare's chart. The full chart has a space for every day from January to September.

Nomination form

The nomination form should be a legal and binding reservation of the owner's mare to a stallion. Failure to do this can only lead to the mare owner being disappointed or annoyed should the stallion's book become full. Some stallion owners may desire to charge a 'booking fee' in advance. The nomination form can vary from stud to stud − an example is shown in Fig. 12.6.

On a busy stud, veterinary visits are frequent, and concise records must be kept. These should include the date the mare has had a visit, the mare owner's name and address, the mare's name and any veterinary treatment. It is useful if these are recorded in a duplicate notebook so that the veterinary surgeon can also keep a copy for his records.

Whenever a mare's feet have been trimmed or shod by the farrier or she has been wormed by a member of staff, this information should be recorded, thus making sure the mare owner will be invoiced.

NOMINATION FORM FOR: Stallion...

Season...................Stud Fee...........+V.A.T.

Mare Owner's Name and Address:

....

....

....

Tel No:...

Mare's Particulars

Name... Age...

Breed... Registration No:...........................

Sire... Dam..

Colour.. Height..

Is mare in foal or barren?...

Or a maiden?...

Date foaled................................... or due to foal...............................

Foal's sire...

Dates last in season...

Or due in season..

Normal length of oestrus........................days in ....days off

Do you agree to promotion of oestrus if necessary?.....................................

Length of stay at stud...

Scan required before departure from stud?...

Mare's Behaviour:

Handling and catching...

with other horses...

at teasing/covering...

Fig. 12.6 Nomination Form. (See page 164 for continuation.) (The 'General Conditions', which make up the first part of the form, have been already shown in Fig. 2.1.) (Courtesy: Broadstone Stud, Oxfordshire.)

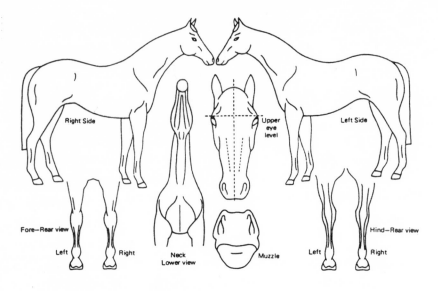

White markings to be shown in ink

Other comments:

....

....

....

I confirm my booking of a nomination to the stallion:

....

I have read and agree with the General Conditions of Broadstone Stud.

Signed.. Date................................

Fig. 12.6 (cont.)

Covering certificates

At the end of the stud season, unless the mare owner particularly requires a certificate to prove that the mare has been covered by a certain stallion, a stallion service certificate and foaling certificate should only be sent to an owner once the invoice has been settled

(Fig. 12.7). Figure 12.8 shows an example of a foal registration certificate.

Mares' foaling dates

A record must also be kept of the dates that mares are expected to foal so that they can be managed accordingly.

Grassland management

Horses require grass paddocks both for nutrition and for exercise and in many cases the requirement for exercise takes precedence over the feeding value of the pasture; consequently, hay or concentrates are fed to make good the shortfall. However, the survival of a stud demands economic use of grassland, particularly if there are large numbers of visiting mares, and the stud manager must be capable of good quality grassland maintenance.

The productivity of a paddock will depend on several factors; altitude, aspect, rainfall, soil type, wind speed, direction and temperature, drainage and grass types.

Soil type

Soils are complex mixtures of minerals, decaying matter, air and water, and their composition can vary considerably. Soil should be tested to ascertain its type and acidity or alkalinity; grass grows best at a pH of 6.5 to 6.8. Sandy soils tend to be too acidic and need the addition of ground limestone to correct the acidity every four years whereas clay soils may only need to be limed every seven years.

Drainage

If a soil is poorly drained it holds too much moisture and is unproductive and poaches badly in the winter. Draining wet soils is very expensive and it is an important aspect of grassland management to maintain existing drainage. Drainage ditches should be

British Warm-Blood Horses Co Ltd t/a

Head Office: Moorlands Farm, New Yatt, Witney, Oxon

Tel: 099 386 673

STALLION SERVICE CERTIFICATE NO

			Cobalt (XX)
		Millerole (XX)	Muscida (XX)
	Dutch Courage (NWP)		Avenir (Geld)
		Higonia (NWP)	Cigonia
DUTCH NIGHT WH76			Abjan
		Emir de la Fot (SF)	Solamide
	Vol de Nuit (BWB)		Commandeur
		Cedola (NWP)	Bedola

This is to certify that the MARE

Reg No   Graded   Breed

Born   Colour   Height

Owned by

Address

         Tel No:

was covered by   on:

1st service   2nd service   3rd service

Service fee: payable on first service

Signature of stallion owner or representative

No

Stallion's name
DUTCH NIGHT WH76

Mare owner

Address

Mare

Reg No   Born

Colour   Height

Breed

Service dates:

1st

2nd

3rd

Date of arrival

Date of departure

Service fee £

Groom's fee £

Keep £

Total £

Date foal born

Sex   Colour

Name

FOALING CERTIFICATE

Please return to:

Mr J Rose
Paradise Farm
South Newington
Oxon

Sire:

Dam:

	Day	Month	Year
Date of			
1st service			
2nd service			

Date foal born

Sex   Colour

Name

Markings:

Head

Neck

Body

Legs: LF

       RF

       LH

       RH

Mare owner

Address

Date

Fig. 12.7 Stallion service certificate and foaling certificate.

FOAL REGISTRATION CERTIFICATE

This form is to be completed by a veterinary surgeon and
forwarded with a cheque for £12.00 to

The British Warm-Blood Society Registration Office
Botterley Hill Cottage
Springes Lane
Faddiley, Nr Nantwich
Cheshire   **Tel: 027 074 417**

not later than 6 months after the birth of the foal.
Late registration (after 6 months) £18.00.
Foals born after 31 December 1985 must be registered within
6 months of birth and cannot apply for late registration.

Proposed name of foal   1st choice

                      2nd choice

Please make sure that the first letter of your foal's name is the same as the
first letter of the sire's name (excluding any prefixes)

Date foal born   Sex   Colour

Head

Neck

Legs: LF

       RF

       LH

       RH

Body

Owner's signature

Owner's name in full

BWBS membership no

Address

Veterinary surgeon's signature

Address

                                 Date

BWBS membership is not compulsory when registering an animal.

Fig. 12.8 Foal registration certificate.

regularly checked for blockages, stagnant water, fallen banks and blocked drain exits, and culverts or pipes taking drainage under gates must be maintained.

Grass types

The horse's hoof is designed to grip by digging into the turf, and horses are notorious for 'poaching' paddocks so that they resemble ploughed fields more than grassland. This means that the sward needs to be tough and resilient with well-established tillering species of grass. The grass production of the sward also needs to be spread over the whole growing season. Both these points mean that modern quick-growing temporary leys containing few species of grass tend to be unsuitable for horse paddocks. Instead, a more old-fashioned type of seed mixture containing both early and late maturing grass species is required. These grasses must be able to withstand the horses' close-biting grazing and send out shoots to produce a dense sward with a good bottom. As it is generally considered undesirable to put large amounts of fertiliser on horse paddocks it is also useful for the seed mixture to contain some clover; clover is a leguminous plant with root nodules which contain nitrogen-producing bacteria.

A basic horse paddock mixture could contain:

(a) two perennial ryegrass varieties (*Lolium perenne*) to make up 50 per cent of the mixture;
(b) two creeping red fescues (*Festuca rubra*) to make up 25 per cent of the mixture;
(c) the remaining 25 per cent being made up of:

   (i) crested dog's tail (*Cynosurus cristatus*) 5–15 per cent;
   (ii) rough or smooth stalked meadow grass 5–15 per cent;
   (iii) wild white clover 1–2 per cent.

Some timothy may also be included if hay is to be taken (see Fig. 12.7 for examples of suitable grasses.)

Characteristics of chosen grass species:

Perennial ryegrass: This forms the basis of many grass seed mixtures as it is very versatile and persistent under all types of conditions. However, it tends to decline after a few years on poor light soils unless well fertilised.

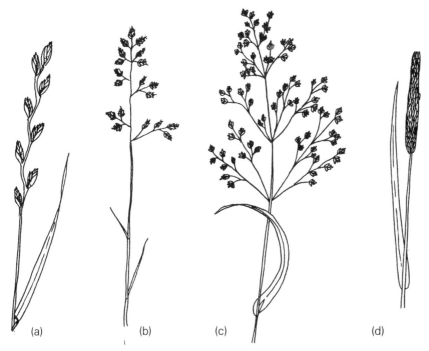

Fig. 12.9 Examples of some grasses suitable as constituents of a basic horse paddock mixture. (a) Rye grass; (b) Red fescue; (c) Rough meadow grass; (d) Timothy.

Creeping red fescue: This grows well in difficult conditions and on hill land and is used extensively for sports turf.

Rough stalked meadow grass: Highly palatable and grows well on moist, rich soils; it gives good 'bottom' to the sward by growing close to the ground and keeping out undesirable species.

Smooth stalked meadow grass (Kentucky blue grass): Able to withstand drought and useful on light dry sandy soils.

Wild white clover: Grows on almost every kind of soil as it has deep roots and can withstand drought. Farm varieties are 'aggressive' and can take over a paddock as they are designed to withstand frequent cutting, so wild types should be selected for grazing paddocks. However, wild types will die out if large amounts of nitrogen fertiliser and heavy grazing are used. Seed companies may market grass seed mixtures specifically for horse paddocks.

Pasture management

It is important that pasture is managed efficiently if the stud is going to be successful; a good grassland management routine is outlined below.

The grassland year:

Autumn: Subject to analysis, add lime every three to seven years (0.5−2.6 tonnes/ha), phosphate every two to three years (80−120 units/ha) and potash every year (60−120 units/ha). The pasture should not be grazed for about three weeks until rain has washed these inorganic fertilisers into the soil.

Spring: The paddocks should be harrowed and if necessary some nitrogen applied. Weed control is best done now before the weeds grow away. The paddocks should be rolled to consolidate the soil and redress the effects of any poaching.

Spring and summer: Ideally, droppings should be picked up; if not, harrowing should take place weekly on a warm, drying day to spread out the droppings and kill worm larvae. However, once-monthly harrowing with the tined side of the harrow followed by rolling may well be adequate.

Any rank grass that grows to 12 inches (30 cm) or more should be topped to 6 or 8 inches (15−20 cm) unless the horses are being followed by cattle and/or sheep. Cattle and sheep tidy up the grass and eat horse worm larvae and can also bring in some income to the stud. Sheep are useful on wet ground as they do not poach the ground as much as cattle. Sheep can also be used to graze a newly planted ley as their close grazing habit encourages tillering without damaging the grass.

Any undesirable weeds which have survived spring weed control should be pulled up and burnt before they seed, or may be spot treated using a knapsack sprayer. There are new regulations that stipulate that anybody doing such spraying must have attended a training course and gained a proficiency certificate.

Winter: Horses should be removed from paddocks as soon as they start to get badly poached. This will obviously depend on the weather and soil type. Horses may not be able to be turned out on clay soils from Christmas until April but a sandy soil may well support limited grazing all winter. It may be necessary to 'write off' one small turn-out paddock every winter and reseed it in the Spring.

If there are too few horses to graze the paddocks effectively then the grass will need to be conserved as hay, silage or haylage. This can be done by a contractor unless the stud is prepared to invest in the necessary machinery.

Appendix 1
Northern and Southern Hemisphere equivalents

As far as possible the seasons of the year rather than months have been used throughout the book; spring is spring wherever you are. Occasionally, however, months have inevitably been used and in order to avoid confusion the following table can be used.

Northern Hemisphere		Southern Hemisphere
Dec		June
Jan	Winter	July
Feb		Aug
Mar		Sept
Apr	Spring	Oct
May		Nov
June		Dec
July	Summer	Jan
Aug		Feb
Sept		Mar
Oct	Autumn	Apr
Nov		May

Appendix 2

The stud year (For Non-Thoroughbred Stud)

Month	Stallions	Mares	Youngstock	Routine treatments
January	Turn lights on. Begin exercise. Increase concentrates.	Bring in mares at night. Lights on for barren mares, rugs and increased concentrates.	In at night.	Worm. Trim feet. Check teeth in need of rasping at least every twelve months.
February	Teasing, swabbing. Exercise.	Barren mares teased and swabbed.		Worm.
March	Teasing, covering. Exercise if not covering.	Foaling.		Worm. Trim feet.
April	Teasing, covering.	Foaling. Scanning.		Worm.
May	Teasing, covering.	Foaling, out at night unless showing. Scanning.	Turned out.	Worm. Trim feet.
June	Teasing, covering.	Foaling. Scanning.		Worm.
July	Teasing, covering.	Scanning.		Check teeth. Worm. Trim feet. Worm.
August	Teasing, covering. Exercise if few mares.			Worm. Trim feet.
September	Rough-off and turn out if possible or begin ridden work.	Weaning – cut food.	Weaning. 3 year olds backed. Colts castrated.	Worm.
October	Rest or ridden.	Weaning – cut food. Pregnancy diagnosis for 1 October terms.	Weaning. In at night. Handling	Worm.
November	Rest or ridden. In at night. Flu and tetanus vaccination.	Feeding at grass. Flu and tetanus vaccination.		Worm. Trim feet.
December	Rest or ridden. In at night.	Feeding at grass.		Worm.

Appendix 3
Summary of stud information

Oestrus:	In heat/in season/in use/on:	Approx. 5 days
Dioestrus:	Not in season/off:	Approx. 17 days
Anoestrus:	Not in season:	Late autumn → early spring

Gestation period: Approx. 340 days
After 325 days described as FULL TERM
Between 300−325 days described as PREMATURE and may die
Before 300 days no chance of survival

Pregnancy diagnosis:	Manual:	42 days to end of pregnancy
	Blood test:	45−90 days
	Urine sample:	100 days onwards
	Scanning:	Approx. 21 days
Foal normals:	Temperature	99°F−101°F (37.3°C−38.3°C)
	Pulse	at birth 80 beats per min. rising to 140 and returning to 100
	Respiration	30 to 40 per minute

Appendix 4
Contagious Equine Metritis and Reproductive Diseases

Contagious Equine Metritis and Reproductive Diseases
Recommended Code of Practice − Revised 1981
Horse Race Betting Levy Board

All mares will require a negative clitoral swab before arrival at stud. Forms A and B should be available before mare arrives.

After arrival at stud

Low risk mares

One negative clitoral swab on arrival, plus one endometrial/cervical swab when in the first oestrus for aerobic culture for venereal pathogenic bacteria (infoal mares at foaling oestrus).

High risk mares

(Previously infected or in contact mares *plus* imported mares other than from France or Ireland)

Mares in foal (to foal at stud)

Negative clitoral swab before arrival.
Negative clitoral swab after arrival.
Negative clitoral and endometrial swab during foaling heat.

High risk barren mares

Negative clitoral swab before going to stud.
Negative clitoral swab after arrival.
Negative clitoral and endometrial swab at first oestrus at stud.

High risk maiden mares

As barren mares above.
Foals from infected mares and/or high risk mares should be swabbed before three months of age.

Stallions and teaser stallions

Two sets of swabs taken at not less than 7 days interval after 1 January 1981. Results entered on Form C, which should be available for mare owners' perusal.

Hygiene

All handlers should be made aware of rules of hygiene and the danger of contaminating mares.
Use tail bandages and disposable gloves when handling mares' and stallions' genitalia.
Isolate all aborting mares and have a full investigation.
Vulval discharges to be investigated and the mare isolated.

THIS CODE APPLIES TO THOROUGHBRED BREEDERS IN UK, FRANCE AND IRELAND.
Other breeders should consult the stud owners and breed societies for their requirements.

Appendix 5
Glossary

Abortion: Expulsion of the products of conception from the uterus between day 30 and day 300 of pregnancy.

Accessory sex glands: Glands of stallion's reproductive tract that contribute secretions to the seminal fluid.

Acrosome: A cap-like membrane which covers the head of the spermatozoon.

Acute: Having short and relatively severe duration.

Afterbirth: The expelled placenta, amnion and cord, normally expelled 10−140 minutes after foal's birth.

Allantochorion: The membrane formed by the fusion of the allantois and chorion.

Allantoic fluid: Brown or yellowish-brown waste fluid, formed partly by placenta and partly by foetal urine; helps to protect foetus and lubricate birth canal.

Allantois: The inner sac of the placenta.

Amino acids: Organic building blocks for protein.

Amnion: The innermost of the membranes covering the embryo. Filled with amniotic fluid in which the embryo is free to move and is cushioned from mechanical injury.

Anoestrus: Period of sexual inactivity, generally occurring in the winter when the mare does not come into season.

Anterior pituitary gland: A gland located at the base of the brain which secretes several hormones including LH and FSH.

Anthelmintic: Substance administered to destroy parasitic larvae and/or eggs.

Antibody: Complex protein molecule which combines with molecules of antigen, thus protecting the animal against disease.

Anus: The external opening of the rectum.

Aorta: The major artery that carries blood away from the heart to the rest of the body.

Artery: Blood vessel that carries blood away from the heart.

Artifical insemination (AI): The process of depositing semen into the female reproductive tract by artificial means.

Artificial vagina (AV): Device used to collect semen for evaluation or AI.

Back-raking: Removal of faeces before rectal examination.

Barker foal: A foal suffering from a convulsive syndrome thought to be caused by lack of oxygen at birth, characterised by convulsions, lack of suck reflex and wandering behaviour.

Barren: A mare (other than a maiden mare) that did not become pregnant during the last breeding season.

Breaking water: Expulsion of allantoic fluid during birth of foal.

Breech birth: Hindquarters of foal presented first.

Broad ligaments: Fibrous bands of tissue which suspend the reproductive tract of the mare from the upper wall of the abdominal cavity.

Bulbourethral glands: Paired accessory sex glands of stallion which add volume and nutrients to the ejaculate.

Caesarian section: Surgical removal of the foetus through an incision in the abdominal and uterine walls.

Carbohydrate: Substance containing carbon, hydrogen and oxygen; includes starch, sugar and cellulose and broken down by the gut to provide energy.

Caslick's operation: Stitching a section of the vulval lips together to prevent air and contaminants from entering the reproductive tract.

Catheter: A hollow cylinder designed for withdrawing fluid from or introducing fluid into the body.

Cervix: The muscular neck-like structure that separates the uterus from the vagina.

Chorion: The outermost of the three placental membranes.

Chorionic villi: Minute projections on the chorion which attach the placenta to the uterus; the sites of gas, nutrient and waste exchange between the dam and foetus.

Chromosome: Protein strands within the cell nucleus which carry genetic information.

Chronic: Long-term, not acute.

Clitoris: A small cylindrical erectile body situated in the lower portion of the vulva.

Colostrum: The first milk of the mare; it provides the newborn foal with protective antibodies and also acts as a laxative.

Conception: Combination of genetic material of sperm and ovum.

Conceptus: The products of conception, i.e. the embryo and foetal membranes.

Congenital: Existing at and usually before birth, referring to conditions that may or may not be inherited.

Contagious equine metritis: Highly contagious disease which causes inflammation of uterine, cervical and vaginal membranes.

Coprophagy: Eating manure.

Corpus haemorrhagicum: The blood filled cavity formed at the site of ovulation immediately after the egg is released.

Corpus luteum: The mass of glandular cells formed after ovulation at the site of ovulation, responsible for prostaglandin production.

Crude protein: Refers to the total nitrogen content of a feed.

Cryptorchid: Horse with one or both testicles retained in body cavity.

Dam: Female parent

Digestible energy: The portion of the energy content of a feed that the animal is able to digest and absorb.

Dioestrus: A period between seasons when the mare is not receptive to the stallion; lasts 15−16 days.

Donor mare: The female that contributes the embryo for embryo transfer.

Ductus arteriosus: A foetal vessel that connects the left pulmonary artery with the descending aorta. Closes after birth.

Dummy foal: See *Barker foal.*

Dystocia: Abnormal or difficult birth.

Ejaculation: Emission of seminal fluid.

Embryo: The conceptus up to 30−40 days gestation.

Embryo transfer: A method whereby an embryo is removed from its natural mother and implanted in the uterus of a host mother.

Endocrine: Of a ductless gland that produces an internal secretion.

Endometrial cups: Raised structures formed on about day 36 of pregnancy in the pregnant uterine horn; produce PMSG.

Endometritis: Inflammation of the endometrium.

Endometrium: The lining of the uterine wall.

Entropian: Infolding of the eyelid.

Epididymis: A U-shaped tubular structure attached to the long axis of each testis, comprised of head, body and tail.

Epiphyseal growth plate: Cartilagenous area that lies between the metaphysis and the epiphysis; site of long bone elongation.

Epiphysis: End of a long bone which is separated from the bone by a plate of cartilage in the immature animal.

Erectile tissue: Tissue containing vascular spaces that become engorged with blood.

Erection: The state of erectile tissue when blood-filled.

Extender: Fluid used for semen dilution.

Fallopian tube: Oviduct; tube leading from ovary to end of uterine horn.

Fertile: Able to produce offspring.

Flagellum: A lash or whip-like appendage.

Flehman's posture: Common behaviour pattern consisting of curling and raising upper lip.

Foal heat: First post-foaling oestrus period; occurs 7–9 days after foaling.

Foal heat scours: Diarrhoea that commonly occurs seven to nine days after foaling.

Foetus: Unborn foal from about day 40 of gestation to birth.

Foley catheter: A catheter equipped with an inflatable cuff used to retain the catheter within the body.

Follicle: A small sac or cavity.

Follicle stimulating hormone (FSH): Anterior pituitary hormone causing follicular growth in the mare and stimulating sperm production in the stallion.

Foramen ovale: An opening in the foetal heart which closes soon after birth.

Full term: Refers to the mature state of the foetus at birth.

Gamete: A sex cell; sperm or ovum.

Gel fraction: The viscous, gelatinous fraction of the ejaculate.

Gene: The functional unit of heredity carried on the chromosomes.

Genital tract: Reproductive tract; reproductive passageway and all associated organs.

Genitals: The organs of reproduction.

Gestation: Pregnancy.

Gland: A secreting organ.

Gonad: An organ that produces gametes, e.g. testis and ovary.

Gonadotrophin: A hormone that affects gonad function, e.g. FSH, LH.

Graafian follicle: A mature ovarian follicle.

Granulosa cell tumour: A tumour of the ovary arising from the membrana granulosa of the Graafian follicle.

Granulosa cells: Cells which line the inside of the ovarian follicle

and support the developing ovum.
Gravid: Pregnant.

Haematoma: A localised mass of blood that is restricted to a definite space or within a tissue.
Haemorrhage: Bleeding.
Herpes virus: A group of viruses, one of which causes equine herpes virus.
Hippomane: Brown, free-floating object found in the placental tissues after birth and thought to be formed from mineral and protein deposits in the allantoic cavity.
Hormone: A chemical which is formed in one organ and carried in blood to another organ to elicit a specific response.
Hypothalamus: An area at the base of the brain which attaches to and controls the pituitary gland.

Immunity: Power to resist infection.
Implantation: Also placentation. Attachment of the placenta to the endometrium between days 45 and 150 of gestation.
Inherent: Natural, inborn.
Intromission: Insertion.
In utero: In the uterus.
In vivo: In the living body.
Involution: The process by which an organ returns to its former size and cellular state, e.g. uterus after pregnancy.

Kicking boots: Padded boots sometimes placed on mares' hind feet to protect the stallion from kicks.

Labia: The vulval lips.
Lactation: The production of milk.
Libido: Sex drive.
Lumen: The cavity within a tube.
Luteinising hormone (LH): A hormone produced by the anterior pituitary that stimulates the development of corpora lutea in the mare and of interstitial tissue in the male.
Luteolytic: Causing degeneration of luteal tissue.
Lysine: An amino acid essential in the diet of the growing horse.

Maiden: A mare that has not been covered.
Mammary gland: The udder, a collection of highly modified oil glands that synthesise and secrete milk.

Mastitis: Inflammation of the mammary gland.

Meconium: The first droppings of a foal.

Metritis: Inflammation of the uterus.

Monorchidism: Complete absence of one or both testes.

Motility: The ability of a sperm cell to move in a normal forward manner.

Mucus: The clear viscous secretion of the mucous membranes consisting of mucin, epidermal cells, leucocytes and salts.

Neonatal maladjustment syndrome: See *Barker foal.*

Neonate: Foal less than four days old.

Nucleus: The body within a cell that encloses the cell's chromosomes.

Oestrogen: A hormone produced by the ovaries which causes a mare to come into season.

Oestrous cycle: The cyclic series of oestrus and dioestrus periods occurring within the breeding season.

Oestrus: Period during which the mare is receptive to the stallion.

Oocyte: Immature ovum.

Oogenesis: A series of cell divisions starting with the primary germ cells in the ovary and resulting in the production of ova.

Ossification: To form bone or change into bone.

Ova: Plural of ovum.

Ovary: One of the paired gonads in the female which contain ova.

Oviduct: See *Fallopian tube.*

Ovulation: The release of an egg from a mature follicle.

Ovulation fossa: An anatomical feature of the equine ovary located on the pinched-in face of the ovary; site at which ovulation occurs.

Ovum: The egg produced in the ovary which contains the mother's genetic information to be passed on to the offspring.

Palpation: Feeling or perceiving by sense of touch.

Parturition: Process of giving birth.

Passive immunity: Disease resistance acquired through transfer of antibodies from another individual, e.g. to foal via colostrum.

Pathogen: A disease-causing organism.

Pelvis: A ring of bones with sacrum above, ilia at sides and pubic and ischial bones below. Bones form area of attachment for muscles and ligaments which control hindquarters.

Penis: Male organ of copulation composed of spongy erectile tissue and containing the urethra.

Perineum: The area between the thighs from the anus to the mammary glands or scrotum.

Pessary: Vaginal suppository.

Phantom mare: A raised, padded support mounted by the stallion during semen collection.

Pipette: A tube used to transport small amounts of a gas or liquid.

Placenta: A vascular organ that surrounds the foetus during gestation. It is connected to the foetus by the umbilical cord and serves as the structure through which the foetus receives nourishment and excretes waste.

Placentation: Attachment of the placenta to the uterine wall.

Postnatal: After birth.

Post partum: Pertaining to, or occurring during, the period following parturition.

Post sperm fraction: Sticky gel secreted by the seminal vesicles.

Pregnant mare serum gonadotrophin (PMSG): Hormone produced by the endometrial cups between days 40 and 180 of gestation; action similar to mixture of LH and FSH.

Premature: Pertaining to a foal born between days 300 and 325 of gestation.

Prenatal: Before birth.

Prepuce: The sheath or fold of skin that covers the glans penis.

Presperm fraction: That fraction of the ejaculate which precedes the sperm-rich fraction, cleaning and lubricating the penis.

Progesterone: Hormone produced by corpus luteum which is essential to maintain pregnancy.

Prolactin: A hormone of the anterior pituitary gland that stimulates the secretion of milk.

Prostaglandin: $PGF_{2\alpha}$, a hormone that causes regression of the corpus luteum in the non-pregnant mare.

Prostate: A gland consisting of two bodies situated either side of the urethra, its secretions are added to the seminal plasma.

Purulent: Consisting of, containing or forming pus.

Pyometra: An accumulation of pus in the uterus.

Raw semen: Semen which has not been treated by filtering or addition of extenders, etc.

Recipient mare: The mare receiving a transferred embryo.

Rectal palpation: Examination of organs or structures adjacent to the rectum by feeling for size, texture, etc.

Rectum: The terminal portion of the alimentary tract.

Resorption: The removal of an exudate, blood, pus, etc., by absorption.

Retained placenta: Afterbirth not expelled 6 to 12 hours after foaling.

Scrotal hernia: Portion of intestine enters scrotum.

Scrotum: Pouch which contains the testicles.

Semen: The ejaculate containing spermatozoa and secretions of testes and accessory glands.

Seminal fluid: See *Semen.*

Seminal vesicles: Pear-shaped accessory sex glands adding fluid and nourishment to the semen.

Seminiferous tubules: Tubules located in the testes in which sperm cell production occurs.

Septicaemia: Systemic disease caused by the presence of pathogenic micro-organisms and their toxic products in the blood.

Sheath: See *Prepuce.*

Silent mare: Mare which ovulates but fails to show behavioural signs of oestrus.

Smegma: Cheesy sebaceous secretion which can accumulate in the male horse's sheath.

Speculum: An instrument for enlarging the opening of a canal or cavity in order to permit examination.

Spermatic cord: The cord-like structure that suspends the testicle within the scrotum and which contains the vas deferens, blood vessels and nerves.

Spermatogenesis: Development of mature sperm cells from primitive germ cells.

Spermatozoon: Male gamete or sex cell, comprised of head, neck and tailpiece.

Spermicide: An agent which kills sperm.

Sterile: Infertile or barren, aseptic.

Stillbirth: Birth of a dead foetus.

Stocks: Restraining crate.

Subfertility: Slight infertility.

Swabbing: Process of obtaining samples of secretions from the vagina, cervix, uterus, clitoral fossa and penis using sterile absorptive material.

Systemic infection: One which invades the whole body.

Teaser: Male horse used to test the mares being in season.

Teat: The nipple of a mammary gland.

Testis: The male gonad where spermatozoa are formed.

Testosterone: Male hormone produced in testes and responsible for development and maintenance of secondary sexual characteristics.

Turgid: Swollen, congested.

Twitch: An instrument used to restrain a horse by placing pressure on the upper lip.

Udder: The mammary gland.

Umbilical cord: Attaches the placenta to foetus and carries urine and blood between the foetus and placenta.

Urachus: Small vessel in the umbilical cord that connects the foetal bladder to the allantois.

Uterine prolapse: Inversion of the uterus, resulting in partial or complete expulsion from the vulva.

Uterus: Hollow muscular organ consisting of cervix, uterine body and uterine horns.

Vagina: The genital passageway that extends from the cervix to the vulva.

Vas deferens: The excretory duct of the testicle.

Vein: Blood vessel that carries blood towards the heart.

Viable: Capable of living.

Vice: An undesirable habit.

Virus: A microscopic agent or infectious disease that can only reproduce in living tissue.

Vulva: The external genitalia of the mare.

Wanderer foal: See *Barker*.

Waxing up: The formation of beads of dried colostrum at the ends of the teats.

Weaning: To permanently deprive foals of their mothers' milk.

Winking: Protrusion of the clitoris between the labia.

Yellow body: See *Corpus luteum*.

Zygote: Fertilised ovum.

Appendix 6
Useful equestrian addresses

European addresses

Ada Cole Memorial Stables Ltd: Mr. E. B. Collier (Director), Broadlands, Broadley Common, Nr. Nazeing, Waltham Abbey, Essex. EN9 2DH. Tel: 099−289−2133

American Saddlebred Association of Great Britain: Mrs. Cheryl Lutring, Uplands, Alfriston, E. Sussex. Tel: (0323) 870295

Ancient Order of Pack Riders: Mrs. C. R. Stone, Fridays Acre, Bromyard Road, Stoke Bliss, Tenbury Wells, Worcestershire. WR15 8RU.

Arab Horse Society: Registrar and Chief Administrator, Goddards Green, Cranbrook, Kent. TN17 3LP. Tel: 0580−713389

Association of British Riding Schools: Miss A. Lawton, Old Brewery Yard, Penzance, Cornwall. TR18 2SL. Tel: (0736) 69440

Association of Show and Agricultural Organisations: Mr. J. N. Armitage, The Showground, Winthorpe, Newark, Notts. NG24 2NY. Tel: Newark 702627

Bransby Home of Rest for Horses: Mr. P. E. Hunt, Bransby, Nr. Saxilby, Lincoln. LN1 2PH.

British Andalusian Society: Mrs. J. Bernard, East Cottage, Skeete Road, Lyminge, Nr. Folkestone, Kent. Tel: (0303) 863178

British Appaloosa Society: Mr. N. Hawkins, c/o 2 Fredrick Street, Rugby. Tel: Rugby 860535/860800

British Bloodstock Agency Ltd: Thormanby House, Falmouth Avenue, Newmarket, Suffolk. CB8 0NB. Tel: Newmarket 665021

British Driving Society: Mrs. J. Dillon, 27 Dugard Place, Barford, Nr. Warwick. CV35 8DX. Tel: 0926−624420

British Equestrian Olympic Fund: Bridget Jennings Promotions, 10 Barleymow Passage, Chiswick, London. W4 4PH. Tel: 01−944−6477

British Equestrian Promotions Ltd: 35 Belgrave Square, London, SWIX 8QB. Tel: 01−235−6431

British Equestrian Trade Association: Wothersome Grange, Bramham, Nr. Wetherby, West Yorks. LS23 6LY. Tel: 0532−892267

British Equine Veterinary Association: Mrs. Carpenter (Secretary), Park Lodge, Bells Yew Green Road, Frant, Tunbridge Wells, Kent. TN3

9EB. Tel: 089–275–368

British Field Sports Society: Gen. J. Hopkinson CB (Director), 59 Kennington Road, London, SE1 7PZ. Tel: 01–928–4742

British Friesian Horse Society: Mr. M. Kiestra, George & Dragon Hall, Mary Place, London. W11.

British Hay and Straw Merchants Association: Mr. F. W. Burton, Hoval House, Orchard Parade, Mutton Lane, Potters Bar, Herts. EN6 3AR. Tel: Potters Bar 42343

British Lipizzaner Horse Society: Mrs. L. Moran, Austan Stud, King's Walden, Hitchin, Herts. SG4 8NJ. Tel: 043–887–754

British Morgan Horse Society: Mrs. A. Cannor-Bulmer, George and Dragon Hall, Mary Place, London W11. Tel: 01–229–8155

British Mule Society: Mrs. L. Travis, Hope Mount Farm, Top of Hope, Alstonefield, Nr. Ashbourne, Derbyshire. Tel: (033527) 353

British Palomino Society: Penrhiwllan, Llandysul, Dyfed. SA44 5NZ (023 975 387)

British Percheron Society: Mrs. A. Neaves, c/o Neaves and Neat, 52A Broad Street, Ely, Cambs. Tel: (0353) 67005

British Quarter Horse Association Ltd: Mr J. Wood-Roberts, 4th Street, NAC, Stoneleigh, Kenilworth, Warks. CV8 2LG. Tel: (0203) 26850

British Show Hack, Cob and Riding Horse Association: Mrs. R. Smith, Rookwood, Packington Park, Meriden, Warwickshire. CV7 7HF. Tel: Meriden (0676) 23535

British Show Jumping Association: Lt. Cmdr. W. B. Jefferis, RN (Retd), British Equestrian Centre, Stoneleigh, Kenilworth, Warwickshire. CV8 2LR. Tel: Coventry 552511

British Show Pony Society: Mrs. J. Toynton, 124 Green End Road, Sawtry, Huntingdon, Cambridgeshire. Tel: Ramsey 831376

British Spotted Pony Society: Mrs. E. Seymour, Wantsley Farm, Broad Windsor, Beaminster, Dorset. DT8 3PT. Tel: Broad Windsor 68462

British Trakehner Association: Miss D. M. Lorch, Buckwood, Fulmer, Buckinghamshire. Tel: Fulmer 2606.

British Veterinary Association: Mr. P. B. Turner, MA, 7 Mansfield Street, Portland Place, London. W1M 0AT. Tel: 01–636–6541

British Warmblood Society: Mrs. D. Wallin, Moorlands Farm, New Yatt, Witney, Oxfordshire. Tel: (0993) 86673

Caspian Pony Stud (U.K.) and Society: Hopstone Lea, Claverley, Salop. Tel: Claverley 206.

Cleveland Bay Horse Society: Mr. J. F. Stephenson, MA, FRICS, York Livestock Centre, Murton, York. YO1 3UF. Tel: York 489731

Clydesdale Horse Society of Great Britain and Ireland: Mr. R. S. Gilmour, 24 Beresford Terrace, Ayr, Ayrshire, Scotland. Tel: 0292 281650

Coloured Horse and Pony Society: Miss P. Sheppard, 15 Welga Road, Welwyn, Herts. Tel: 043 871 6613

Commons, Open Spaces and Footpaths Preservation Society: Mr. P. Clayden and Mr. D. Mackay, 25a Bell Street, Henley-on-Thames, Oxon. RG9 2BA. Tel: Henley 3535

Connemara Pony Breeders Society: Mrs. P. Macdermott, 73 Dalysfort Road, Salthill, Galway, Ireland. Tel: 091-22909

Dales Pony Society: Miss P. A. Fitzgerald, 55 Cromwell Street, Walkley, Sheffield. S6 5RN. S. Yorks. Tel: Sheffield (0742) 336762

Dartmoor Pony Society: Mrs. E. C. M. Williamson, Weston Manor, Corscombe, Dorchester, Dorset. DT2 0PB. Tel: Corscombe 466.

Donkey Breed Society: Mr. D. J. Demus, Manor Cottage, South Thurseby, Nr. Alford, Lincs. LN13 0AS. Tel: 05216 320

Donkey Sanctuary and International Donkey Protection Trust: Slade House Farm, Salcombe Regis, Sidmouth, Devon. EX10 0NU. Tel: Sidmouth 6391

Endurance Horse & Pony Society of G.B.: Mrs. P. Hancox, 15 Newport Drive, Alcester, Warks. Tel: (0789) 762963

English Connemara Pony Society: Mrs. M. V. Newman, 2 The Leys, Salford, Chipping Norton, Oxon. OX7 5FD. Tel: 0608 3309

Equine Behaviour Study Circle: Ms. G. C. Cooper, 9 Rostherne Avenue, Lowton St. Lukes, Warrington. WA3 2QD. Tel: (0942) 712552

Equine Research Station: Balaton Lodge, Newmarket, Suffolk. Tel: 0638-661111/3009

Eriskay Pony Society: 114 Braid Road, Edinburgh.

Exmoor Pony Society: Mr. D. Mansell, Glen Fern, Waddicombe, Dulverton, Somerset. TA22 9RY. Tel: (03984) 490

Fallabella Society: Lady Rosamund Fisher, Kilverstone Wildlife Park, Thetford, Norfolk. Tel: Thetford (0842) 5369

Farriers' Registration Council: P.O. Box 49, East of England Showground, Peterborough. PE2 0GU. Tel: (0733) 234451

Federation Equestre Internationale: Mr. F. O. Widmer, Schosshaldenstrasse 32, CH-3000 Berne 32, Switzerland. Telex: 32710 fei ch.

Fell Pony Society: Mr. C. Richardson, 19 Dragley Beck, Ulverston, Cumbria. LA12 0HD. Tel: 0229 52742

Fjord Horse Society of Great Britain: Miss L. Moran, Ausdan Stud, Kings Walden, Hitchin, Herts. SG4 8NJ. Tel: Whitwell (043 887) 754

Hackney Horse Society: Miss S. Oliver, 34 Stockton, Warminster, Wiltshire. Tel: 0985-50906

Haflinger Society of Great Britain: Mrs. Robins, 13 Park Field, Pucklechurch, Bristol. BS17 3NS. Tel: 027582 3479

Hanoverian Society: Please refer to Hanoverian Society in Germany or contact: British Warmblood Society.

Heavy Horse Preservation Society: Captain J. K. Murphy and Mr. R. G. Hooper, Old Rectory, Whitchurch, Shropshire. SY13 1LF.

Highland Pony Society: Mrs. S. Bell, Orwell House, Milnathort, Kinrosshire. KY13 7YQ. Tel: (0577) 63495

Home of Rest for Horses: Brig. J. Spurry, CB, MRCVS, Westcroft Stables, Speen Farm, Aylesbury, Buckinghamshire. HP17 0PP. Tel: Hampton Row 464.

Horse Race Betting Levy Board: Mr. R. T. Ricketts, 17/23 Southampton Row, London WC2. Tel: 01–405–5346/6456

Horses and Ponies Protection Association: Greenbank Farm, Greenbank Drive, Fence, Nr. Burnley, Lancashire. BB12 9QJ. Tel: Nelson 65909

Hunters' Improvement and National Light Horse Breeding Society: Mr. G. W. Evans, 96 High Street, Edenbridge, Kent. Tel: (0732) 866277

Hurlingham Polo Association: Col. Harper, Ambersham Stables, Ambersham, Nr. Midhurst, W. Sussex. Tel: Lodsworth 254.

Icelandic Horse: Mrs. J. Elias, Rosebank, Higher Merley Lane, Corfe Mallen, Dorset. BH21 3EG.

International League for the Protection of Horses: 67a Camden High Street, London NW1. Tel: 01–388–1449.

Irish Draught Horse Society of Great Britain: 4th Street, N.A.C., Stoneleigh, Kenilworth, Warwickshire. Tel: Coventry 26850

Irish Horse Board: The Irish Farm Centre (3rd Floor), Bluebell, Dublin 12. Tel: (0001) 501166

Jockey Club: Registry Office, 42 Portman Square, London W1H 0EN. Tel: 01–486–4921.

Ladies Side Saddle Association: Mrs. Bacon, Marches Equestrian Centre, Harewood End, Ross-on-Wye. Tel: 098–987–234

Lipizzaner Society of Great Britain: Starrock Stud, Ludwell, Wilts. SP7 0PW. Tel: (074788) 639

London Harness Horse Parade Society: Mrs. A. Vincent, Young and Co's Brewery PLC, Ram Brewery, Wandsworth High Street, London. SW18 4JD. Tel: 01–870–0141

Lusitano Breed Society: Mrs. L. McCurley, Fox Croft, Bulstrode Lane, Felden, Hemel Hempstead, Herts. HP3 0BP. Tel: 0442 50806

Master of Foxhounds Association: Mr. A. H. B. Hart, Parsloes Cottage, Bagendon, Cirencester, Gloucestershire. Tel: North Cerney 470

Master Saddlers Association: Mr. H. C. Knight, Easdon, Lower Icknield Way, Chinnor, Oxon. OX9 4DZ. Tel. (0844) 52860

National Equine Welfare Committee: c/o Bransby Home of Rest for Horses, Bransby, Nr. Saxilby, Lincoln. LN1 2PH.

National Foaling Bank: Meretown Stud, Newport, Shropshire. Tel: Newport 811234.

National Horse Racing Museum: 99 High Street, Newmarket, Suffolk. Tel: Newmarket 667333

National Master Farriers', Blacksmiths' and Agricultural Engineers' Association: Mr. D. Height, Avenue R, 7th Street, N.A.C., Stoneleigh, Kenilworth, Warks. Tel: (0203) 20870

National Pony Society: Col. A. R. Whent, Brook House, 25 High Street, Alton, Hants. GU34 1AW. Tel: Alton 88333

National Stud: Newmarket, Suffolk. Tel: 0638−663464

National Trainers' Federation: Col. R. J. Mackaness, 42 Portman Square, London. W1H 0EN. Tel: 01−935 2055

New Forest Pony and Cattle Breeding Society: Miss D. Macnair, Beacon Corner, Burley, Ringwood, Hants. BH24 4EW. Tel: Burley 2272

People's Dispensary for Sick Animals (P.D.S.A.): P.D.S.A. House, 21−37 South Street, Dorking, Surrey. RH4 2LB. Tel: Dorking 81691.

Ponies of Britain: Mrs. M. Mills, Chesham House, Green End Road, Sawtry, Huntingdon, Cambs. PE17 5UY. Tel: (0487) 830278

Racehorse Owners' Association: Mr. J. S. Biggs, 42 Portman Square, London. W1H 9FF. Tel: 01−486−6977.

Rare Breeds Survival Trust Ltd: 4th Avenue, N.A.C., Stoneleigh, Nr. Kenilworth, Warwickshire. Tel: Coventry 51141

Riding for the Disabled Association: Miss C. Haynes, Avenue 'R', National Agricultural Centre, Stoneleigh, Kenilworth, Warwickshire. CV8 2LY. Tel: Coventry 56107.

Royal Agricultural Society of England: Mr. A. D. Callaghan, National Agricultural Centre, Stoneleigh, Kenilworth, Warwickshire. CV8 2LZ. Tel: Coventry 696969.

Royal College of Veterinary Surgeons: 32 Belgrave Square, London. SW1 8QP. Tel: 01−235−4971.

Shetland Pony Stud-Book Society: Mr. D. M. Patterson, 8 Whinfield Road, Montrose, Angus. DD10 8SA. Tel: Montrose 73148.

Shire Horse Society: Mr. R. W. Bird, MBE, East of England Showground, Peterborough, PE2 0XE. Tel: Peterborough 234451.

Society of Master Saddlers: Mr. H. C. Knight, Easdon, Lower Icknield Way, Chinnor, Oxon. OX9 4DZ. Tel: (0844) 52860

Suffolk Horse Society: Mr. P. Ryder-Davies, 6 Church Street, Woodbridge, Suffolk. Tel: Wickham Market 746534

Thoroughbred Breeders' Association: Mr. S. G. Sheppard, 168 High Street, Newmarket, Suffolk CB8 9AQ. Tel: 0638−661321.

Veterinary Defence Society: 14 Princess Street, Knutsford. WA16 6BY. Tel: (0565) 52737

Weatherby and Son: Mr. S. M. Weatherby, 42 Portman Square, London. W1N 0EN. Tel: 01−486−4921

Welsh Pony and Cob Society: Mr. T. E. Roberts, 6 Chalybeate Street, Aberystwyth, Dyfed. SY23 1HS. Tel: Aberystwyth 617501.

Western Horseman's Association of Great Britain: Mr. C. Ward, 36 Old Fold View, Barnet, Herts.

World Arabian Horse Organisation: Ms. K. Powell, Calgarth Hall, Windermere, Cumbria. LA23 1HZ. Tel: (09662) 2999

Worshipful Company of Farriers: Mr. F. W. Birch, 3 Hamilton Road,

Cockfosters, Barnet, Hertfordshire. EN4 9EH. Tel: 01—449—5491
Worshipful Company of Loriners: 50 Cheyney Avenue, London. E18
2DR. Tel: 01—989—0652
Worshipful Company of Saddlers: The Clerk, Saddlers' Hall, Gutter
Lane, Cheapside, London. EC2V 6BR. Tel: 01—726—8661/6

American addresses

American Association of Breeders of Holsteiner Horses, Locksley, Mill-
wood, VA 22646.
American Connemara Pony Society, Hoshiekon Farm, PO Box 513,
Goshen, CT 06756.
American Fox Trotting Horse Breeders Association, Inc., PO Box 666,
Marshfield, MO 65706.
American Horse Council, 1700 K Street NW, Suite 300, Washington DC,
20006.
American Morgan Horse Association, Inc., PO Box 1, Westmorland, NY
13490.
American Quarter Horse Association, 2701 I—40 East, Amarillo, TX
79168.
American Saddlebred Horse Association, 4093 Ironworks Pike, Lexington,
KY 40511.
American Shire Horse Association, 1687 NE 56 Street, Altoona, IA 50009.
American Trakehner Association Inc., 5008 Pine Creek Drive, Suite B,
Westerville, OH 43081.
American Warmblood Society, Route 5, Box 1219A, Phoenix, AZ 85009.
Arabian Horse Registry of America, Inc., 12000 Zuni Street, Westminster,
CO 80234.
Cleveland Bay Horse Society of America, Rockbridge Farm, Natural
Bridge, VA 24578.
Clydesdale Breeders of the USA, 17378 Kelley Road, Pecatonica, IL
61063.
Haflinger Association of America, 2624 Bexley Park Road, Bexley, OH
43209.
Hanoverian Society Inc., 831 Bay Avenue, Suite 2—E, Capitola, CA
95010.
National Show Horse Registry, 10401 Linn Station Road, Suite 237,
Louisville, KY 40223.
Palomino Horse Association of America, PO Box 324, Jefferson City, MO
65102.
Percheron Horse Association of America, PO Box 141, Fredericktown.
OH 43019.

The Jockey Club (Thoroughbred), 380, Madison Avenue, New York, NY 10017.

United States Trotting Association (Standard Bred), 750 Michigan Avenue, Columbus, OH 43215.

Warmblood Studbook of the Netherlands, North American Department, 6208 Reba Drive, Roseville, CA 95678.

Welsh Pony Society of America, PO Box 2977, Winchester, VA 22601.

Index